DATE DUE		

THE FACTS ON FILE
GEOMETRY
HANDBOOK

CATHERINE A. GORINI, Ph.D.
Maharishi University of Management, Fairfield, Iowa

☑®
Facts On File, Inc.

To Roy Lane for introducing me to the fascination and joys of geometry
To my parents for their love, support, and encouragement
To Maharishi Mahesh Yogi for the gift of pure knowledge

The Facts On File Geometry Handbook
Copyright © 2003 by Catherine A. Gorini, Ph.D.

Facts On File, Inc.
132 West 31st Street
New York NY 10001

Library of Congress Cataloging-in-Publication Data

Gorini, Catherine A.
 Facts on file geometry handbook / Catherine A. Gorini
  p. cm.
Includes bibliographical references and index.
 ISBN 0-8160-4875-4
 1. Geometry. I. Title: Geometry handbook. II. Title.
QA445.2 .G67 2003
516—dc21   2002012343

Facts On File books are available at special discounts when purchased in bulk quantities for businesses, associations, institutions, or sales promotions. Please call our Special Sales Department in New York at 212/967-8800 or 800/322-8755.

You can find Facts On File on the World Wide Web at
http://www.factsonfile.com

Cover design by Cathy Rincon
Illustrations by David Hodges

Printed in the United States of America

VB Hermitage 10 9 8 7 6 5 4 3 2

This book is printed on acid-free paper.

CONTENTS

ACKNOWLEDGMENTS

When asked by the emperor Ptolemy if there were not a shorter road to geometry than through the *Elements,* Euclid replied that there was "no royal road to geometry." This is indeed a good fortune for all, because it has meant that the one common road to geometry is crowded with students, teachers, amateurs, dilettantes, experts, and nobility, all working together, sharing ideas, making contributions large and small. I have enjoyed all my journeys on this road and would like to acknowledge the fellow travelers who have made this road feel royal to me, particularly Ed Floyd, David Henderson, Walter Meyer, Doris Schattschneider, Daniel Schmachtenberger, Marjorie Senechal, and Bob Stong.

This book could not have been written without the help of many individuals, only a few of whom I can mention here. Great appreciation goes to all the ladies of Maharishi Spiritual University of America at Heavenly Mountain, North Carolina, particularly Meg Custer and Marcia Murphy, for providing a quiet and sublimely blissful atmosphere in which to work. I am indebted to Michelle Allen, Jim Bates, Lynn Ellis, Amit Hooda, Janet Kernis, Abbie Smith, Joe Tarver, and Swati Vivek for their help with library research and graphics. Elizabeth Frost-Knappman and Jim Karpen played an essential role in transforming a seed of an idea into a reality, and for this I am deeply grateful. Frank K. Darmstadt, Senior Editor at Facts On File, provided crucial guidance and all manner of expertise with unfailing patience.

INTRODUCTION

Geometry is the study of shape, form, and space, and is of interest for its own sake as well as for its many applications in the sciences and the arts. Geometry is an integral part of mathematics, drawing upon all other areas in its development, and in turn contributing to the development of other parts of mathematics.

The roots of geometry lie in many different cultures, including the ancient Vedic, Egyptian, Babylonian, Chinese, and Greek civilizations. The development of geometry as a deductive science began with Thales and reached maturity with the *Elements* of Euclid around 300 B.C.E.

In the *Elements,* Euclid sets out a list of statements called postulates and common notions, which are fundamental, self-evident truths of geometry. These statements express common experiences and intuitions about space: It is flat, it extends infinitely in all directions, and it has everywhere the same structure.

From the postulates and common notions, Euclid derives 465 propositions that form the body of Euclidean geometry. Euclid's presentation is systematic, beginning with simple statements that depend only on the postulates and common notions and concluding with intricate and exquisite propositions. Each proposition is supported by a logical deduction or proof, based only on the postulates, the common notions, and other propositions that have already been proved.

Because of its clarity and logical rigor, the axiomatic method of Euclidean geometry soon became the model for how a discipline of knowledge should be organized and developed. The theorems of Euclidean geometry came to be regarded as absolute truth and assumed a special role as the crowning achievement of human thought. Trying to emulate the success of Euclid, philosophers and scientists alike sought to find the self-evident principles from which their disciplines could be derived.

During the 19th century, however, the discovery of non-Euclidean geometries brought an end to the special and unique role of Euclidean geometry. Nicolai Lobachevsky, János Bolyai, and Bernhard Riemann showed that there are other geometries, just as systematic and just as rigorous as Euclidean geometry, but with surprisingly different conclusions.

These new geometries revolutionized the way mathematicians viewed mathematics and the nature of mathematical truth. A mathematical theory must be based on a collection of postulates or axioms, which are not self-evident truths, but instead are rules assumed to be true only in the context of a specific theory. Thus, the axioms of Euclidean geometry are different from those of elliptic geometry or hyperbolic geometry. A mathematical theorem, proven by logical deduction, is not true, but rather only valid, and even then valid only in the realm governed by the axioms from which it was derived. Different axioms give rise to different, even contradictory, theorems.

The discovery of non-Euclidean geometries strengthened the study of geometry, giving it new vigor and vitality and opening new avenues of investigation. However, non-Euclidean geometries were only one aspect of the development of geometry in the 19th century, which was truly a golden age for geometry, with projective geometry, affine geometry, vector spaces, and topology all emerging as important and independent branches of geometry.

Today, after more gradual growth during the first part of the 20th century, geometry is once again flourishing in a new golden age.

There are many factors supporting this renewed interest in geometry. Computers have become both an inspiration and tool for geometry, responsible for computational geometry, computer graphics, and robotics. Many newer areas of geometry, such as combinatorial geometry, discrete geometry, differential geometry, algebraic topology, dynamical systems and fractals, graph theory, and knot theory are the natural unfoldment of discoveries of earlier centuries. Other areas of current interest have their origin in applications: crystallography, frameworks, minimal surfaces, sphere packings, and the string theories of modern physics.

All these areas and more are included in this handbook, which will serve students of geometry, beginner or advanced, and all those who encounter geometric ideas in their pursuit of the sciences, arts, technology, or other areas of mathematics.

Euclidean geometry, trigonometry, projective geometry, analytic geometry, non-Euclidean geometry, vectors, differential geometry, topology, computational geometry, combinatorial geometry, knot theory, and graph theory are just some of the branches of geometry to be found here. The selection of topics has focused on what a student would first encounter in any of these areas. Terms from other parts of mathematics that are needed to understand geometric terms are included in the glossary, but the reader is referred to *The Facts On File Algebra Handbook* and *The Facts On File Calculus Handbook* for a more comprehensive discussion of these topics. Nowhere is it assumed here that the reader has studied calculus.

GLOSSARY

Concrete or abstract, every mathematical concept needs a precise name that can be used to distinguish it from everything else. The great variety of technical terms used in geometry reflects the long history and organic growth of geometry.

Many ordinary words, such as side, face, edge, and foot, have precise geometrical meanings. Still other words, such as segment, graph, and pole, have taken on more than one mathematical meaning. Fortunately, the meanings of such words can be determined from the context in which they appear.

Often, new concepts are named after individuals. Many of these, such as the Euler line, Mandelbrot set, and Penrose tiles, are named for the person who discovered or first studied them. On the other hand, a surprising number of discoveries, including the Simson line and the Reuleaux triangle, are named for individuals who had very little or even nothing to do with them.

Many geometric objects or concepts that were invented in modern times have newly devised names based on Greek or Latin roots. Words such as fractal, polyomino, and polytope have such an origin.

Still other geometric creations were given colorful names based on some distinctive property or feature of their appearance; horseshoe map, monkey saddle, batman, kissing number, tangle, and jitterbug are good examples of this.

There are close to 3,000 definitions given here, from abscissa to zonohedron. Each term is defined as simply as possible and illustrations are used throughout to clarify meanings. Cross-references are given in small capital letters.

BIOGRAPHIES

Geometry today is the result of the work of many individuals, each making contributions great or small, some remembered and some virtually forgotten. The towering greats of mathematics, Isaac Newton, Leonhard Euler, Carl Friedrich Gauss, and David Hilbert, contributed to many areas of mathematics in addition to geometry. There are other great figures, such as Archimedes, Nicolai Lobachevsky, Felix Klein, and H. S. M. Coxeter, whose work was primarily in geometry. Other mathematicians, such as Maria Agnesi, Jean Robert Argand, Giovanni Ceva, and Helge von Koch, are remembered primarily for a single concept bearing their name.

Many discoveries in geometry were the by-products of work in other areas. Leonardo da Vinci and M. C. Escher were artists, Joseph Plateau was a physicist, Pierre Bezier was an engineer, and Edward Lorenz is a meteorologist, yet all made significant contributions to geometry. Still others, such as Henri

Brocard and Robert Ammann, pursued geometry more as a hobby than as a profession, but nevertheless made significant contributions to geometry that now bear their names.

All these and more, for a total of more than 300 individuals, have brief biographies here. Included are more than 60 current figures who are now alive and still contributing to the development of geometry.

CHRONOLOGY

Geometry has had a long and rich development from ancient to modern times, continually acquiring new knowledge without relinquishing past glories. In fact, achievements of the past are inevitably reworked into new results. For example, the straight lines of Euclid have become the geodesics of differential geometry, and Euclid's parallel lines have given rise to the ultraparallels of hyperbolic geometry and parallel transport in differential geometry.

Old problems continue to inspire mathematicians. In recent decades, mathematicians have found solutions to problems first posed by Kepler, Euler, Tait, and Poincaré. Each new discovery reveals more territory to explore, and geometry seems to progress at a faster and faster rate each decade.

More than 200 important milestones in the long history of geometry are listed in this chronology. The list spans more than 4,000 years, with more than a quarter of the achievements since 1950.

CHARTS AND TABLES

The results of geometry are recorded in formulas and theorems, numbers and symbols. In this section, the reader will find more than 100 fundamental theorems from all areas of geometry and detailed instructions for more than 50 compass-and-straightedge constructions. There are also tables of geometric formulas, trigonometry tables, the values of important constants, and the meanings of commonly used symbols.

RECOMMENDED READING

There is a wealth of material available to anyone who wants to explore geometry further. The list given here includes many resources, among them books at an introductory level, resource materials for the classroom teacher, software and websites, and carefully selected advanced books for those who want to probe geometry more deeply.

Each book has its own special angle and flavor, and the full range of geometry, ancient to modern, theoretical to applied, is covered. There are books here to appeal to every taste, need, and interest—everyone can find something suitable for continuing the study of geometry.

SECTION ONE
GLOSSARY

AA   *See* ANGLE-ANGLE.

AAS   *See* ANGLE-ANGLE-SIDE.

Abelian group A GROUP with a COMMUTATIVE BINARY OPERATION.

abscissa The first, or *x*-, coordinate of a point.

absolute geometry Geometry based on all the Euclidean axioms except the fifth, or parallel, postulate.

absolute polarity Any elliptic POLARITY in a PROJECTIVE GEOMETRY that is kept fixed can be used to define an ELLIPTIC GEOMETRY based on the projective geometry. Such an elliptic polarity is called the absolute polarity of the elliptic geometry.

absolute value The absolute value of a number gives its distance from 0. The absolute value of a real number *a* is the greater of *a* and –*a*, denoted |*a*|. Thus |3| = |–3| = 3. The absolute value of a complex number *a* + *bi* is |*a* + *bi*| = $\sqrt{a^2 + b^2}$.

acceleration A measure of how fast speed or velocity is changing with respect to time. It is given in units of distance per time squared.

accumulation point *See* LIMIT POINT.

ace   One of the seven different VERTEX NEIGHBORHOODS that can occur in a PENROSE TILING.

achiral Having REFLECTION SYMMETRY.

acnode An ISOLATED POINT of a curve.

acute angle An angle with measure less than 90°.

acute-angled triangle A triangle whose angles are all acute.

acute golden triangle An isosceles triangle with base angles equal to 72°.

ad quadratum square A square whose vertices are the midpoints of the sides of a larger square.

Adam's circle For a triangle, the Adam's circle is the circle passing through the six points of intersection of the sides of the triangle with the lines through its Gergonne point that are parallel to the sides of its Gergonne triangle.

adequacy *See* COMPLETENESS.

adjacency matrix An *n* × *n* MATRIX that represents a GRAPH with *n* vertices. The entry in the *i*th row and the *j*th column is the number of edges between the *j*th vertex and the *j*th vertex of the graph. In the adjacency matrix for a DIGRAPH, the entry in the *i*th row and the *j*th column is the number of edges from the *i*th vertex to the *j*th vertex.

adjacent Next to. Two angles of a polygon are adjacent if they share a common side, two sides of a polygon are adjacent if they share a common vertex, and two faces of a polyhedron are adjacent if they share a common edge. A point is adjacent to a set if every NEIGHBORHOOD of the point contains some element of the set.

affine basis A set of AFFINELY INDEPENDENT vectors whose AFFINE COMBINATIONS form an AFFINE SPACE.

affine collineation *See* AFFINE TRANSFORMATION.

affine combination A sum of scalar multiples of one or more vectors where the sum of the scalars is 1.

affine coordinates Coordinates with respect to axes that have unrelated units of measurement.

affine geometry The study of properties of INCIDENCE and PARALLELISM, in the EUCLIDEAN PLANE or some other AFFINE PLANE.

affine hull The smallest AFFINE SUBSPACE containing a given set of points.

affine plane A PROJECTIVE PLANE from which the IDEAL LINE has been removed. The EUCLIDEAN PLANE is an example of an affine plane.

affine ratio The ratio *AB/BC* for three collinear points *A, B,* and *C.* This ratio is preserved by AFFINE TRANSFORMATIONS.

affine reflection A STRAIN that maps points not on the fixed line of the strain to the opposite side of the fixed line.

affine set *See* LINEAR SET.

affine space A PROJECTIVE SPACE from which the HYPERPLANE AT INFINITY has been removed.

affine subspace A subspace of an AFFINE PLANE or AFFINE SPACE.

affine transformation A TRANSFORMATION of an AFFINE SPACE that preserves collinearity.

affinely dependent A set of vectors is affinely dependent if there is an AFFINE COMBINATION of them with nonzero coefficients that is the zero vector.

affinely independent A set of vectors is affinely independent if an AFFINE COMBINATION of them is the zero vector only when all the coefficients are 0.

affinely regular polygon A polygon in the AFFINE PLANE whose vertices are images of one another under a given AFFINE TRANSFORMATION.

affinity *See* AFFINE TRANSFORMATION.

Alexander horned sphere

air speed The speed of an object, such as a bird or airplane, relative to the air.

Alexander horned sphere A surface that is topologically equivalent to a sphere but whose complement in three-dimensional space is not SIMPLY CONNECTED.

Alexander polynomial A polynomial determined from the sequence of CROSSINGS in a KNOT or LINK. It is a KNOT INVARIANT.

algebraic curve A curve that is the graph of a polynomial equation or a system of polynomial equations.

algebraic expression An expression built up out of numbers and variables using the operations of addition, subtraction, multiplication, division, raising to a power, and taking a root. The powers and roots used to form an algebraic expression must be integral.

algebraic function A function given by an algebraic expression.

algebraic geometry The study of algebraic equations and their solutions using the geometric properties of their graphs in a coordinate space.

algebraic multiplicity The algebraic multiplicity of an EIGENVALUE λ is the DEGREE of λ as a root of a CHARACTERISTIC POLYNOMIAL.

algebraic surface A SURFACE defined by an ALGEBRAIC FUNCTION.

algebraic topology The study of algebraic structures, such as the FUNDAMENTAL GROUP, associated to TOPOLOGICAL SPACES.

alternate exterior angles Angles that are outside two parallel lines and on opposite sides of a TRANSVERSAL crossing the two lines.

alternate interior angles Angles that are between two parallel lines and on opposite sides of a TRANSVERSAL crossing the two lines.

alternate method A method of constructing a new GEODESIC POLYHEDRON from a geodesic polyhedron. TRIANGULATE each face of the polyhedron, subdivide each triangle into n^2 congruent triangles, project each vertex to the sphere circumscribed about the polyhedron from the center of the sphere, and connect the projected vertices. The number n is the frequency of the geodesic polyhedron thus constructed.

alternate vertices Vertices of a polygon separated by two adjacent sides.

alternating Describing a KNOT or LINK DIAGRAM where CROSSINGS alternate between over and under as one traces the knot. The TREFOIL KNOT is an example of an alternating knot.

alternating angles *See* ALTERNATE INTERIOR ANGLES.

alternating group The GROUP of EVEN PERMUTATIONS. The alternating group is a subgroup of the SYMMETRIC GROUP and is denoted A_n.

alternating prism A TWISTED q-PRISM that has alternate corners from the top and bottom bases TRUNCATED.

alternation *See* DISJUNCTION.

altitude (1) A perpendicular segment connecting a vertex of a polygon to its base or the extension of the base. For a cone or pyramid, the altitude is a perpendicular dropped from the apex to the base. (2) A SMOOTH FUNCTION defined on the surface of a sphere that is positive and BOUNDED.

ambient isotopy For two SUBSETS A and B of a TOPOLOGICAL SPACE S, a DEFORMATION of S that MAPS A to B.

ambiguous case In constructing a triangle from given data, the ambiguous case occurs when two sides of a triangle and an angle opposite one of the sides are given. It is ambiguous because there can be two noncongruent triangles that satisfy the given conditions.

Ammann bar A segment marked on the tiles of an APERIODIC TILE to be used as a guide for matching tiles. The Ammann bars form lines on a tiling of tiles that have been marked in this way.

amphicheiral knot An oriented KNOT equivalent to its MIRROR IMAGE (a positive amphicheiral knot) or the REVERSE of its mirror image (a negative amphicheiral knot).

amplitude (1) *See* ARGUMENT OF A COMPLEX NUMBER. (2) *See* POLAR ANGLE.

analysis The study of the theoretical foundations of CALCULUS and its generalizations.

analysis situs A name for topology used in the 19th century.

analytic geometry Geometry that makes uses of numerical coordinates to represent points. Analytic geometry usually refers to the use of the Cartesian plane, but can refer to the use of other coordinate systems as well.

analytic proof An algebraic proof, usually using a coordinate system, of a geometric property.

anchor ring *See* TORUS.

angle A planar figure formed by two rays with a common endpoint. The two rays are called the sides of the angle and their common endpoint is called the vertex of the angle. The interior of an angle is one of the two regions in the plane determined by the two rays that form the angle. The measure of an angle is determined by that part of a circle that one ray sweeps out as it moves through the interior of the angle to reach the other ray. Often, two segments are used to represent an angle. The measure of an angle is usually represented by a lowercase Greek letter.

angle between two curves The angle formed by a tangent line to one of the curves and a tangent line to the other curve at a point of intersection of the two curves.

angle bisector A ray that divides an angle into two congruent angles.

angle of a polygon The interior angle formed by two adjacent sides of a polygon.

angle of depression For a viewer looking at an object below the horizon, the angle between a ray from the viewpoint to the horizon and a ray from the viewpoint to the object viewed.

angle of elevation For a viewer looking at an object above the horizon, the angle between a ray from the viewpoint to the horizon and a ray from the viewpoint to the object viewed.

angle of parallelism In HYPERBOLIC GEOMETRY, the angle of parallelism for a line parallel to a given line through a given point is the angle between the parallel line and a perpendicular dropped from the given point to the given line.

angle of rotation The angle through which a given figure or pattern is rotated about a given center.

angle of sight The smallest angle, with vertex at the viewer's eye, that completely includes an object being observed.

angle preserving *See* CONFORMAL.

angle sum The sum of the measures of the interior angles of a polygon.

angle-angle If two angles of one triangle are congruent to two angles of another triangle, the triangles are similar, and the ratio of proportionality is equal to the ratio of any pair of corresponding sides.

angle-angle-side If two angles of one triangle are congruent to two angles of another triangle, the two triangles are similar, and the ratio of proportionality is equal to the ratio of the given sides, which are adjacent to the second given angle of each triangle.

angle-regular polygon A polygon with all angles congruent to one another. For example, a rectangle and a square are both angle-regular.

angle-side-angle If two angles of one triangle are congruent to the angles of another triangle, the triangles are similar and the ratio of proportionality is equal to the ratio of the included sides.

angular defect In HYPERBOLIC GEOMETRY, the sum of the measures of the three angles of a triangle subtracted from 180°. The area of the triangle is a multiple of its angular defect.

angular deficiency At a vertex of a polyhedron, 360° minus the sum of the measures of the face angles at that vertex.

angular deficit *See* ANGULAR DEFICIENCY.

angular deviation The measure of the angle with vertex at the origin and with sides connecting the origin to any two points in the Cartesian plane.

angular distance The angle between the lines of sight to two objects of observation with vertex at the eye of the viewer.

angular excess In ELLIPTIC GEOMETRY, the sum of the three angles of a triangle minus 180°. The area of the triangle is a multiple of its angular excess.

angular region All points in the interior of an angle.

anharmonic ratio *See* CROSS RATIO.

anisohedral polygon A polygon that admits a MONOHEDRAL TILING of the plane but does not admit any ISOHEDRAL TILINGS.

annulus The region between two concentric circles.

anomaly *See* POLAR ANGLE.

antecedent The first term of a ratio. The antecedent of the ratio $a{:}b$ is a.

anticlastic A saddle-shaped surface.

anticommutative A binary operation represented by $*$ is anticommutative if $a * b = -b * a$. For example, subtraction is anticommutative.

antihomography A product of an odd number of INVERSIONS.

antiparallel Two lines are antiparallel with respect to a transversal if the interior angles on the same side of the transversal are equal. A segment with endpoints on two sides of a triangle is antiparallel to the third side if it and the third side are antiparallel with respect to each of the other two sides. Two lines are antiparallel with respect to an angle if they are antiparallel with respect to the angle bisector. The opposite sides of a quadrilateral that can be inscribed in a circle are antiparallel with respect to the angle formed by the other two sides.

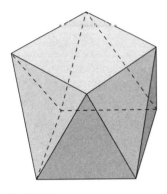

Antiprism

antiparallel vectors Vectors that point in opposite directions.

antipodal Two points are antipodal if they are the endpoints of a diameter of a circle or sphere. The antipodal mapping of a circle or sphere takes a point to its antipodal point.

antiprism A polyhedron with two congruent parallel faces (the bases of the antiprism) joined by congruent isosceles triangular faces (the lateral faces of the antiprism).

antiradical axis For two nonconcentric circles, the antiradical axis is the locus of the center of a circle that intersects each of the given circles at diametrically opposite points. It is parallel to the RADICAL AXIS of the two circles.

antisimilitude *See* INDIRECT ISOMETRY.

antisnowflake A fractal curve formed by replacing each edge of an equilateral triangle by four congruent edges ⌢ pointing toward the center of the triangle and repeating this process infinitely.

antisphere *See* PSEUDOSPHERE.

antisquare curve A fractal curve formed by replacing each edge of a square by five congruent edges ⌐⌐ pointing toward the center of the square and repeating this process infinitely.

antisymmetric relation An antisymmetric relation is a RELATION R with the property that if $a \, \text{R} \, b$ and $b \, \text{R} \, a$ are both true, then $a = b$. For example, the relations $\leq$ and $\geq$ are both antisymmetric.

antisymmetry A COLOR SYMMETRY of a two-color design that interchanges the two colors. Sometimes, an OPPOSITE SYMMETRY.

antitrigonometric function An INVERSE trigonometric function.

antiversion INVERSION followed by a rotation about the center of the circle of inversion.

Antoine's necklace A fractal formed by replacing a TORUS by eight tori linked in a necklace, then replacing each of these tori by eight linked tori, and so on infinitely.

apeirogon A degenerate polygon having infinitely many sides. It consists of a sequence of infinitely many segments on a line and is the limit of a sequence of polygons with more and more sides.

aperiodic tiling A tiling which has no TRANSLATION symmetries.

apex (1) The vertex of an isosceles triangle that is between the two equal sides. (2) The vertex of a cone or pyramid.

Apollonian packing of circles A PACKING by circles that are tangent to their neighbors.

Apollonius, circle of The set of all points such that the distances to two fixed points have a constant ratio.

Apollonius, problem of The problem of constructing a circle tangent to three given circles.

apothem A perpendicular segment connecting the center of a regular polygon to the midpoint of one of its sides.

apotome A segment whose length is the difference between two INCOMMENSURABLE numbers. An apotome has irrational length.

application of areas The use of rectangles to represent the product of two numbers.

arbelos A CONCAVE region bounded by three semicircles, the smaller two of which are contained in the largest. The diameters of the two smaller semicircles lie on the diameter of the largest semicircle and the sum of the two smaller diameters is equal to the larger diameter.

arc (1) The portion between two points on a curve or between two points on the circumference of a circle. (2) An edge of a graph.

arc length A measure of the distance along a curve between two points on the curve.

arccos *See* INVERSE COSINE.

arccot *See* INVERSE COTANGENT.

arccsc *See* INVERSE COSECANT.

Archimedean coloring A COLORING of a tiling in which each vertex is surrounded by the same arrangement of colored tiles.

Archimedean polyhedron *See* SEMIREGULAR POLYHEDRON.

Archimedean property *See* ARCHIMEDES, AXIOM OF.

Archimedean space A space satisfying the AXIOM OF ARCHIMEDES.

Archimedean spiral A spiral traced out by a point rotating about a fixed point at a constant angular speed while simultaneously moving away from the fixed point at a constant speed. It is given in polar coordinates by $r = a\theta$, where a is a positive constant.

Archimedean tiling *See* SEMIREGULAR TILING.

Archimedean value of π The value of π determined by ARCHIMEDES, 3 1/7.

Archimedes, axiom of For any two segments, some multiple of the smaller segment is longer than the larger segment.

Archimedes, problem of The problem of dividing a sphere into two SEGMENTS whose volumes have a given ratio.

arcsec *See* INVERSE SECANT.

arcsin *See* INVERSE SINE.

arctan *See* INVERSE TANGENT.

arcwise connected A region is arcwise connected if every pair of points in the region can be connected by an arc that is completely contained in the region.

area A measure of the size of a two-dimensional shape or surface.

area of attraction of infinity *See* ESCAPE SET.

area-preserving mapping A function that preserves the area enclosed by every closed figure in its domain.

areal coordinates Normalized BARYCENTRIC COORDINATES; i.e., barycentric coordinates in which the sum of the coordinates for any point is 1.

arg *See* ARGUMENT.

Argand diagram *See* COMPLEX PLANE.

argument The independent variable of a function or a value of the independent variable, especially for a trigonometric function.

argument of a complex number The value of θ in the interval $[0°, 360°)$ for a complex number expressed in polar form as $r(\cos \theta + i \sin \theta)$. It is the measure of the directed angle from the positive real axis to a ray from the origin to the graph of the number in the complex plane.

arithmetic-geometric mean The arithmetic-geometric mean of two numbers a and b is obtained by forming two sequences of numbers, $a_0 = a$, $a_1 = \frac{1}{2}(a + b)$, $a_2 = \frac{1}{2}(a_1 + b_1)$, . . ., $a_{n+1} = \frac{1}{2}(a_n + b_n)$, . . . and $b_0 = b$, $b_1 = \sqrt{ab}$, $b_2 = \sqrt{a_1 b_1}$, . . ., $b_{n+1} = \sqrt{a_n b_n}$. Eventually, a_n will equal b_n and that value is the arithmetic-geometric mean of a and b, denoted M(a, b).

arithmetic geometry The study of the solutions of systems of polynomial equations over the integers, rationals, or other sets of numbers using methods from algebra and geometry.

arithmetic mean For two numbers a and b, the arithmetic mean is $(a + b)/2$. For n numbers $a_1, a_2, \ldots, a_n$, the arithmetic mean is $(a_1 + a_2 + \ldots + a_n)/n$.

arithmetic sequence An infinite sequence of the form $a, (a+r), (a+2r), \ldots$.

arithmetic series An infinite sum of the form $a + (a+r) + (a+2r) + \ldots$.

arm A side of a right triangle other than the hypotenuse.

armillary sphere A model of the celestial sphere. It has rings showing the positions of important circles on the celestial sphere.

arrangement of lines A collection of lines in a plane which partition the plane into convex regions or cells. An arrangement is simple if no two lines are parallel and no three lines are concurrent.

artichoke A type of DECAPOD.

ASA *See* ANGLE-SIDE-ANGLE.

ascending slope line *See* SLOPE LINE.

asterix A type of DECAPOD.

astroid The EPICYCLOID traced by a point on the circumference of a circle rolling on the outside of a fixed circle with radius four times as large as the rolling circle. It has four CUSPS.

astrolabe A mechanical device used to measure the inclination of a star or other object of observation.

astronomical triangle A triangle on the celestial sphere whose vertices are an object being observed, the zenith, and the nearer celestial pole.

asymmetric unit *See* FUNDAMENTAL DOMAIN.

asymptote A straight line that gets closer and closer to a curve as one goes out further and further along the curve.

asymptotic Euclidean construction A compass and straightedge construction that requires an infinite number of steps.

asymptotic triangle In HYPERBOLIC GEOMETRY, a triangle whose sides are two parallels and a transversal. An asymptotic triangle has just two vertices.

attractive fixed point A FIXED POINT of a DYNAMICAL SYSTEM that is also an ATTRACTOR.

attractive periodic point A PERIODIC POINT of a DYNAMICAL SYSTEM that is also an ATTRACTOR.

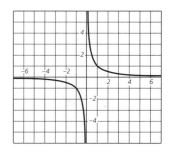

The *x*- and *y*-axes are asymptotes of the hyperbola $y = \frac{1}{x}$.

attractor A point or set with the property that nearby points are mapped closer and closer to it by a DYNAMICAL SYSTEM.

augmented line A line together with an IDEAL POINT.

augmented plane A plane together with an IDEAL LINE.

augmented space Three-dimensional space together with an IDEAL PLANE.

automorphism An ISOMORPHISM from a set to itself.

autonomous dynamical system A DYNAMICAL SYSTEM that is governed by rules that do not change over time.

auxiliary circle of an ellipse The circumcircle of the ellipse; it is the circle whose center is the center of the ellipse and whose radius is the semimajor axis of the ellipse.

auxiliary lines Any lines, rays, or segments made in a construction that are necessary to complete the construction but are not part of the final construction. Also, any lines, rays, or segments drawn in a figure to help prove a theorem.

auxiliary triangle A triangle, usually a right triangle, that can be constructed immediately from the givens in a construction problem.

auxiliary view A two-dimensional projection of a three-dimensional object. Generally, three auxiliary views are used to portray a three-dimensional object.

average curvature For an arc of a curve, the total curvature divided by the arc length. It is measured in degrees or radians per length.

axial collineation A COLLINEATION that leaves each point of some given line fixed.

axial pencil The set of all planes through a given line.

axiom A statement giving a property of an UNDEFINED TERM or a relationship between undefined terms. The axioms of a specific mathematical theory govern the behavior of the undefined terms in that theory; they are assumed to be true and cannot be proved.

Axiom of Choice The statement that a choice can be made of one element from each set in a collection of sets. The Axiom of Choice is usually included as one of the axioms of set theory and is used mainly for infinite collections of infinite sets.

axiomatic method The use of AXIOMATIC SYSTEMS in mathematics.

axiomatic system A systematic and sequential way of organizing a mathematical theory. An axiomatic system consists of UNDEFINED

TERMS, AXIOMS, DEFINITIONS, THEOREMS, and PROOFS. The undefined terms are the fundamental objects of the theory. The axioms give the rules governing the behavior of the undefined terms. Definitions give the theory new concepts and terms based on the undefined terms and previously defined terms. Theorems are statements giving properties of and relationships among the terms of the theory and proofs validate the theorems by logical arguments based on the axioms and previously established theorems. All modern mathematical theories are formulated as axiomatic systems.

axis A line that has a special or unique role. For example, a line used to measure coordinates in analytic geometry is a coordinate axis.

axis of a range *See* BASE OF A RANGE.

axis of curvature *See* POLAR AXIS FOR A POINT ON A CURVE.

axis of homology *See* AXIS OF PROJECTION.

axis of perspectivity The line of intersection of a plane and its image plane with respect to a PERSPECTIVITY. It is left pointwise fixed by the perspectivity.

axis of projection The line containing the intersections of the CROSS JOINS of pairs of corresponding points of a PROJECTIVITY between two lines.

axis of similitude A line containing three or more CENTERS OF SIMILITUDE for three or more circles or spheres.

axonometric projection A projection of a three-dimensional object onto a plane.

azimuth (1) The measure of an angle between the direction of true north and the line of sight to an observed point. (2) The polar angle in a cylindrical or spherical coordinate system.

azimuthal projection A projection from a sphere to a tangent plane from a point that is located on the diameter that meets the tangent plane. An azimuthal projection preserves the direction from the center point to any other point.

balanced coloring A coloring of a band ornament or tiling such that each color occurs equally often.

ball The interior of a sphere (open ball) or a sphere together with its interior (closed ball).

band ornament A pattern or design on an infinite strip whose SYMMETRY GROUP includes TRANSLATIONS in the direction of the strip.

Baravalle spiral A spiral formed by shaded regions of a sequence of nested AD QUADRATUM SQUARES.

barycenter *See* CENTROID.

barycentric coordinates HOMOGENEOUS COORDINATES with respect to two fixed points on a line, three fixed points on a plane, four fixed points in space, and so on. The barycentric coordinates of a point *P* tell what masses, which may be negative, must be placed at the fixed points so that the point *P* is the center of mass of the system.

barycentric subdivision A TRIANGULATION of a polygon formed by connecting the BARYCENTER to each vertex of the polygon. To get a finer subdivision, this process may be repeated, giving the second barycentric subdivision, and so on.

base In general, a base is a side of a polygon or face of a solid to which an altitude is dropped.

base angle An angle adjacent to the base of a polygon.

base of a cone The flat planar region that is part of the surface of a finite cone.

base of a polygon A side of the polygon.

base of a range The line containing the points of a range of points.

base of a Saccheri quadrilateral The side of a SACCHERI QUADRILATERAL between the two right angles.

base of a trapezoid Either of the two parallel sides of a TRAPEZOID.

base of an isosceles triangle The side of an isosceles triangle between the two congruent sides.

base point of a loop The starting and ending point of a LOOP.

base point of a space A fixed point of a topological space that is used as a reference point.

base space of a fiber bundle *See* FIBER BUNDLE.

baseline In HYPERBOLIC GEOMETRY, a line perpendicular to all the lines in a pencil of ULTRAPARALLELS.

basic parallelepiped A PRIMITIVE CELL that is a PARALLELEPIPED.

basic parallelogram A PRIMITIVE CELL that is a PARALLELOGRAM.

basin The set of all points that get closer and closer to an ATTRACTOR of a DYNAMICAL SYSTEM.

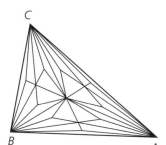

The third barycentric subdivision of triangle *ABC*

basis   A LINEARLY INDEPENDENT SPANNING SET for a space.

basis of a pencil of points   The line containing all the points of the pencil.

basis of a topology   A collection of open sets whose unions and finite intersections are all the open sets of the TOPOLOGY.

bat   A HEXIAMOND shaped like a bat.

batman   A type of DECAPOD.

batter   Architectural term for the reciprocal of a slope; it is the run over the rise.

bearing   A fixed compass direction, usually given in degrees measured from north.

beetle   A type of DECAPOD.

Beltrami sphere   *See* PSEUDOSPHERE.

Beltrami-Klein model   A model of HYPERBOLIC GEOMETRY consisting of the interior of a circle; the lines of the geometry are chords of the circle without their endpoints.

bend   A measure of how each circle fits into a diagram of four circles that are mutually tangent at six distinct points. For a circle in such a diagram, the bend is the reciprocal of its radius, but multiplied by -1 only if the circle contains the other three.

bending   A transformation of a surface that leaves arc length and angle measure invariant. For example, there is a bending from a plane to a cylinder.

Betti number   For a topological space, a number that measures the number of "holes" in each dimension. For example, the Betti numbers for a sphere are 1 in dimension 2 and 0 in dimension 1.

between   A relation among three distinct COLLINEAR points. The point B is between A and C if B is incident to the segment AC. In some axiomatic systems, between is an undefined term.

Bézier spline   A polynomial curve that passes through specified points with specified tangents at each of those points. Bézier splines are used by computer graphics programs to draw a smooth curve through a given set of points with given directions at each point.

biangle   *See* DIGON.

bicentric polygon   A polygon which has both a CIRCUMCIRCLE and an INCIRCLE. Every triangle is a bicentric polygon.

biconditional   *See* EQUIVALENCE.

bicontinuous transformation *See* HOMEOMORPHISM.

bifurcation A qualitative change in the behavior of a DYNAMICAL SYSTEM effected by a small change in the values of the PARAMETERS defining the system.

bijection A function that is both INJECTIVE and SURJECTIVE.

bijective Both INJECTIVE and SURJECTIVE.

bilateral A DIGON whose two vertices are ANTIPODAL.

bilateral symmetry REFLECTION SYMMETRY in the plane. Usually, a shape is said to have bilateral symmetry if it has only one mirror line.

bilinear map *See* MÖBIUS TRANSFORMATION.

bimedian A segment joining the midpoints of opposite sides of a quadrilateral or the midpoints of opposite edges of a tetrahedron.

binary operation A rule that assigns one element of a set to each ordered pair of elements in the set. For example, addition and subtraction are binary operations.

Bing link An UNLINK with two component circles and four crossings. Each circle crosses over the other circle twice.

binomial segment A segment whose length is the sum of two INCOMMENSURABLE numbers. A binomial segment has irrational length.

binormal indicatrix The image of a space curve on a unit sphere where the image of a point on the curve is the tip of the UNIT BINORMAL VECTOR to the curve at that point displaced so its tail is at the center of the unit sphere.

binormal line A line that is a NORMAL of a curve in three-dimensional space and is perpendicular to the PRINCIPAL NORMAL at a given point.

binormal vector A vector that is a NORMAL VECTOR of a curve in three-dimensional space and is perpendicular to the PRINCIPAL NORMAL at a given point.

bipartite graph A GRAPH in which the vertices can be separated into two sets; each edge of the graph connects a vertex in one set to a vertex in the other.

birectangular Having two right angles.

bisect To divide into two congruent pieces.

bitangent A line or plane tangent to a curve or surface at two distinct points.

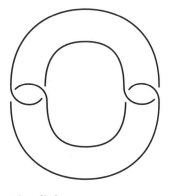

Bing link

blocking segment A segment on a VISIBILITY MAP that separates a VISIBLE REGION closer to the viewpoint from a region that is invisible further away.

body-centered cubic lattice The BODY-CENTERED LATTICE formed by adding the points at the center of each cube in a CUBIC LATTICE.

body-centered cubic packing A PACKING of three-dimensional space by congruent spheres in which the centers of the spheres are at the points of a BODY-CENTERED CUBIC LATTICE. Each sphere is tangent to six other spheres.

body-centered lattice A LATTICE consisting of the points of a given lattice together with the center points of each PRIMITIVE CELL of the lattice.

Borromean rings A LINK consisting of three circles such that if any one is removed, the other two can be separated.

bound vector A vector whose INITIAL POINT is fixed.

boundary The points on the edge of a set. The boundary of a set S contains each point such that every open ball containing the point contains points in the set S and points not in the set S.

bounded function A function that never gets larger than some given maximum value and never gets smaller than some given minimum value.

bounded set A set that can be contained by a circle or sphere of finite radius.

bouquet The figure formed by a finite number of circles or spheres having one point in common.

bowtie A PATCH of a PENROSE TILING that looks like a bowtie. Bowties come in short and long versions.

box *See* RECTANGULAR PARALLELEPIPED.

Boy's surface A one-sided NONORIENTABLE surface that cannot be embedded in three-dimensional space without self-intersections.

brace A rod or segment added to a FRAMEWORK, usually with the intention of making the framework RIGID.

brace graph A GRAPH of a square GRID together with diagonals showing the placement of braces.

braced grid A square GRID together with diagonal BRACES connecting opposite vertices of some of the squares.

braid A finite collection of disjoint vertical curves or strands in three-dimensional space. The strands may weave over and under each other. Two braids are equivalent if one can be DEFORMED into the other while keeping the endpoints of the strands fixed.

braid group A GROUP whose elements are BRAIDS with a fixed number of strands. The group operation is defined by joining the endpoints at the top of one braid to the endpoints at the bottom of the other braid.

braid index For a KNOT, the least number of strands in a BRAID that can be transformed into the knot by joining endpoints at the top of the braid to corresponding endpoints at the bottom of the braid.

branch An EDGE of a GRAPH.

Bravais lattice A three-dimensional LATTICE.

Brianchon point The point of CONCURRENCY of the three diagonals connecting opposite vertices of a hexagon circumscribed about a circle or other conic section.

Brianchon-Pascal configuration A configuration of nine points and nine lines or segments, with three points on each line and three lines passing through each point.

bride's chair A name for the figure used by Euclid in his proof of the Pythagorean theorem.

bridge A connected segment of a KNOT DIAGRAM that crosses over one or more other segments of the diagram.

bridge number The number of BRIDGES in a KNOT DIAGRAM.

brightness For a solid, the area of any one of its ORTHOGONAL PARALLEL PROJECTIONS onto a plane.

Brocard angle An angle whose vertex is at the vertex of a triangle with one side lying on a side of the triangle and the other side a ray from the vertex to the first or second BROCARD POINT of the triangle.

Brocard circle The circle passing through the SYMMEDIAN POINT, the two BROCARD POINTS, and the CIRCUMCENTER of a triangle.

Brocard point A Brocard point is the intersection of three circles, each of which is tangent to one side of a triangle and has another side of the triangle as a chord. Each triangle has two Brocard points. The first or positive Brocard point is obtained by taking the same counterclockwise sequence for each circle; thus, for triangle *ABC* with vertices labeled in the counterclockwise direction, the first circle is tangent to *AB* and passes through *B* and *C*, the second circle is tangent to *BC* and passes through *C* and *A*, and the third circle is tangent to *CA* and passes through *A* and *B*. The second or negative Brocard point is determined by going in the clockwise direction when constructing the circles.

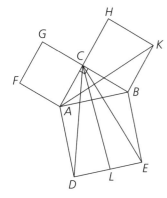

Bride's chair

Brocard ray A ray from a vertex of a triangle to either of its BROCARD POINTS.

broken line A sequence of segments such that consecutive segments share an endpoint.

Brunnian link A LINK consisting of three or more circles that cannot be separated; however, if any one circle is removed, the remaining circles can be separated.

buckyball A molecular form of carbon discovered in 1985. Each molecule has 60 atoms of carbon located at the vertices of a truncated ICOSAHEDRON.

bundle The set of all lines and planes passing through a given point.

bundle of circles The family of circles for a fixed point, called the radical center of the bundle, such that the point has the same power with respect to each circle in the family.

butterfly (1) A STRANGE ATTRACTOR shaped like a butterfly, discovered by Edward Lorenz in his study of long-range weather prediction. (2) A HEXIAMOND shaped like a butterfly.

butterfly effect A characterization of the extreme sensitivity that a DYNAMICAL SYSTEM can have with respect to its INITIAL CONDITIONS; the term was coined by meteorologist Edward Lorenz, who said, "Does the flap of a butterfly's wings in Brazil set off a tornado in Texas?"

Cabri Computer software that allows the user to create, manipulate, and measure geometric figures. It was awarded the Apple Trophy for educational software in 1988 and is available on many platforms.

CAD/CAM Computer Aided Design and Computer Aided Manufacture. CAD/CAM software makes extensive use of geometry to assist in the design of buildings and manufacture of machine parts.

cage *See n*-CAGE.

Cairo tessellation A tessellation of the plane by congruent convex equilateral pentagons that have two nonadjacent right angles; so called because it can be found on streets in Cairo.

calculus The study of changing quantities and their relationships to one another.

calculus of variations The use of CALCULUS to find functions that satisfy specific minimal conditions. For example, the brachistochrone problem can be solved using the calculus of variations.

CAM *See* CAD/CAM.

canal surface A surface that is the ENVELOPE of a family of spheres whose centers lie on a given curve.

cancellation law The law that $ab = 0$ implies $a = 0$ or $b = 0$. It is true for multiplication of real numbers but not MATRIX multiplication.

canonical Standard or usual. Used in a variety of contexts to indicate an obvious or conventional choice.

canonical parallel projection A PARALLEL PROJECTION with the image plane equal to the xy-plane in three-dimensional coordinate space and having direction of projection parallel to the z-axis.

canonical perspective projection A PERSPECTIVE PROJECTION with the image plane equal to the xy-plane in three-dimensional coordinate space and with center of projection on the z-axis.

Cantor discontinuum *See* CANTOR SET.

Cantor fractal dust *See* CANTOR SET.

Cantor set A fractal formed from a segment by removing the middle third without its endpoints, and then removing the middle thirds without their endpoints of the resulting two segments, and so on. The result is an infinite set of points.

Cantor's Axiom For every infinite SEQUENCE of segments such that each segment contains the next, there is a point contained in every segment in the sequence.

cap A region on a sphere whose boundary is a circle on the sphere.

cardioid The heart-shaped EPICYCLOID traced when the fixed circle and rolling circle are congruent. It is given in polar coordinates by $r = a(1 - \cos\theta)$ where a is a constant.

carom To rebound after an impact. When an object caroms off a curve, the angle of impact is equal to the angle of rebound.

Cartesian coordinates The coordinates of a point with respect to perpendicular axes.

Cartesian lattice *See* STANDARD LATTICE.

Cartesian oval The locus of points whose distances r_1 and r_2 from two fixed points satisfy the equation $r_1 + m\,r_2 = a$, for constant and a. The conic sections are special cases of the Cartesian oval.

Cartesian plane A coordinate plane with two perpendicular real coordinate axes.

Cartesian product For two sets, S and T, the set $S \times T = \{(s, t) \,|\, s \in S \text{ and } t \in T\}$. For example, the Cartesian product of two lines is a plane and the Cartesian product of two circles is a torus.

cartwheel A Penrose tiling or a patch of a Penrose tiling that has fivefold rotational symmetry.

case (1) One specific possibility of a more general situation or condition. For example, given two segments, there are two cases: one in which the segments are congruent and one in which they are not congruent. (2) The smallest CONVEX solid that contains a given polyhedron.

Cassini oval The locus of a point the product of whose distances to two fixed points is constant. It is given by the equation $((x - 2)^2 + y^2)((x + a^2) + y^2) = k^4$. If $k = a$ in this equation, the curve is called the LEMNISCATE OF BERNOULLI.

catacaustic *See* CAUSTIC CURVE.

Catalan solid A DUAL of any one of the SEMIREGULAR POLYHEDRA.

catastrophe An abrupt change in a DYNAMICAL SYSTEM brought about by a smooth change in the values of the PARAMETERS governing the system. Such a change is a singularity of the function defining the dynamical system.

catastrophe point A point at which an abrupt change in a DYNAMICAL SYSTEM occurs; SINGULARITY.

catastrophe theory The study of the GLOBAL behavior of functions in terms of properties of their SINGULARITIES.

categorical Describing an AXIOMATIC SYSTEM for which all MODELS are ISOMORPHIC. Euclidean geometry is categorical while ABSOLUTE GEOMETRY is not.

catenary The curve that a hanging chain naturally assumes. Its equation is $y = \cosh x$.

catenoid The surface of revolution created when a catenary is rotated about its axis.

cathetus A side of a right triangle other than the hypotenuse.

caustic curve The curve that is the envelope of light rays emitted from a single point source that have been either reflected (catacaustic) or refracted (diacaustic) by a given curve.

Cavalieri's principle The principle that two solids have equal volumes if every plane parallel to a given plane intersects the solids in planar sections with equal areas.

Cayley line A line passing through three KIRKMAN POINTS and one STEINER POINT. Any hexagon inscribed in a conic section has 20 Cayley lines.

Cayley numbers *See* OCTONIANS.

cell   A polyhedron that is a part of the boundary of a higher-dimensional POLYTOPE. It is the higher-dimensional analogue of the face of a polyhedron. The following are specific important examples:

> 5-cell   A four-dimensional polytope with five vertices and five cells, each a tetrahedron.

> 8-cell   *See* HYPERCUBE.

> 16-cell   A four-dimensional polytope with eight vertices and 16 cells, each a tetrahedron.

> 24-cell   A four-dimensional polytope with 24 vertices and 24 cells, each an octahedron.

> 120-cell   A four-dimensional polytope with 600 vertices and 120 cells, each a dodecahedron.

> 600-cell   A four-dimensional polytope with 120 vertices and 600 cells, each a tetrahedron.

cell-regular Having regular cells or regular solid faces.

cellular automaton A grid of cells that evolve according to rules based on the states of surrounding cells.

center   In general, the point in the middle of a geometric figure.

center of a parallelogram The point of intersection of the two diagonals of a parallelogram.

center of a pencil of lines The point common to all the lines of a PENCIL.

center of a regular polygon The point in a regular polygon that is equidistant from the vertices.

center of antiparallel medians This is an older term used to refer to the symmedian point. *See* SYMMEDIAN POINT.

center of curvature The center of a CIRCLE OF CURVATURE.

center of gravity The point at which a physical object can be balanced. For a geometric shape, it is the point at which a physical model of the shape could be balanced.

center of inversion The center of a CIRCLE OF INVERSION.

center of perspectivity *See* PERSPECTIVITY.

center of projection *See* PERSPECTIVE PROJECTION.

centesimal angle *See* GON.

centesimal minute A unit of angle measure; it is 1/100 of a GON.

centesimal second A unit of angle measure; it is 1/100 of a CENTESIMAL MINUTE.

centigon *See* CENTESIMAL MINUTE.

central angle An angle formed by two radii of the same circle.

central collineation *See* PERSPECTIVE COLLINEATION.

central conic An ellipse or hyperbola; so called because they both have point symmetry.

central dilation *See* DILATION.

central inversion *See* POINT SYMMETRY.

central projection The projection of a sphere from its center onto a tangent plane.

central reflection *See* POINT SYMMETRY.

central symmetry This term is used in different contexts to refer to a dilation or a point symmetry. *See* DILATION and POINT SYMMETRY.

central vanishing point The point on the HORIZON LINE in a perspective picture where the images of lines perpendicular to the picture plane meet. It is not necessarily in the center of the picture.

centrode The centrode of two curves is the locus of the INSTANTANEOUS CENTER OF ROTATION of a rigid body that has a point fixed on each curve.

centroid The CENTER OF GRAVITY of a geometric shape. For a triangle, it is the point of intersection of the three medians.

cevian A segment from a vertex of a triangle to a point on the opposite side or its extension.

cevian triangle A triangle whose vertices are the feet of three concurrent cevians of a given triangle.

chain A sequence of edges of a GRAPH such that two consecutive edges share a vertex.

chaos The extreme sensitivity to INITIAL CONDITIONS seen in some DYNAMICAL SYSTEMS where a very small change in the initial condition produces a dramatic change in the long-term behavior of the system.

characteristic determinant For a square matrix A, the DETERMINANT of $A - \lambda I$.

characteristic equation The equation $\det(A - \lambda I) = 0$ for a square matrix A. Its roots are the EIGENVALUES of the LINEAR TRANSFORMATION A.

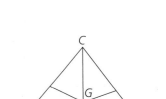

Point G is the centroid of triangle ABC

characteristic function For a given subset T of a set S, the function that has value 1 on elements of T and value 0 on all other elements of S.

characteristic polynomial *See* CHARACTERISTIC DETERMINANT.

characteristic value *See* EIGENVALUE.

characteristic vector *See* EIGENVECTOR.

chart *See* COORDINATE NET.

chickens A version of the PENROSE TILES in which the kite and dart tiles are replaced by tiles that look like chickens.

chiliagon A regular polygon with 1,000 sides. It was studied by ARCHIMEDES.

chiral Not having REFLECTION SYMMETRY. For example, the letter P is chiral.

chord A line segment whose two endpoints lie on a circle or other curve.

chordal distance The distance between the projections onto the RIEMANN SPHERE of two points in the complex plane. The chordal distance between two complex numbers z_1 and z_2 is

$$(z_1, z_2) = \frac{2|z_1 - z_1|}{\sqrt{(1 + |z_1|^2)(1 + |z_2|^2)}}.$$

choropleth A shaded map showing which regions are visible and which regions are invisible from a given VIEWPOINT.

chromatic coloring *See* MAP COLORING.

Cinderella Java-based computer software that allows the user to create, manipulate, and measure geometric figures. It received the digita[2001] award for educational software.

circle The set of all points in a plane at a given distance from a fixed point which is called the center of the circle.

circle geometry The study of properties of figures that are invariant under INVERSION.

circle group The GROUP of ROTATIONS of a circle.

circle of antisimilitude *See* MIDCIRCLE.

circle of curvature For a point on a curve, the circle that best approximates the curve at that point. Its radius is the radius of curvature at that point and its center lies on the principal normal.

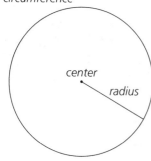

circumference

center

radius

Circle

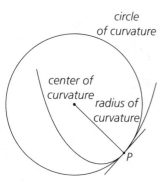

circle
of curvature

center of
curvature
radius of
curvature

P

Circle of curvature at point P

circle of inversion The circle left fixed by an INVERSION.

circle packing A PACKING of the plane by circles.

circle-preserving transformation A TRANSFORMATION such that the image of a circle or line is a circle or line.

circuit A path in a GRAPH whose starting vertex is the same as its ending vertex.

circular cone A cone whose base is a circle.

circular cylinder A cylinder whose base is a circle.

circular reasoning A logical fallacy in which one assumes what one is trying to prove.

circular saw A type of DECAPOD.

circular transformation *See* MÖBIUS TRANSFORMATION.

circumcenter The center of a CIRCUMCIRCLE or CIRCUMSPHERE.

circumcircle A circle that intersects every vertex of a polygon. The center of a circumcircle of a triangle is the intersection of the three perpendicular bisectors of the sides of the triangle.

circumference The points of a circle. Also, the measure of the total arc length of a circle; it is 2π times the radius of the circle.

circumparameter A measure of the size of the sets in a collection of sets. It is a number U such that each set in the collection is contained in a ball of radius U.

circumradius The radius of a CIRCUMCIRCLE or CIRCUMSPHERE.

circumscribed circle *See* CIRCUMCIRCLE.

circumscribed cone of a pyramid A cone that has the same vertex as the pyramid and has as its base a circle that is circumscribed about the base of the pyramid.

circumsphere A sphere that intersects every vertex of a polyhedron or polytope. Not every polyhedron has a circumsphere.

cissoid The curve whose polar equation is $r = 2 \sin \theta \tan \theta$. It is also called the cissoid of DIOCLES after its inventor.

cleaver A segment starting at the midpoint of a side of a triangle that bisects the perimeter of the triangle.

clique A set of pairwise adjacent vertices of a GRAPH.

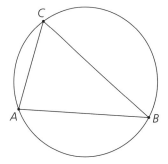

The circumcircle of triangle *ABC*

closed ball A sphere together with the points in its interior.

closed curve A curve, such as a circle, that has no endpoints.

closed disc A circle together with the points in its interior.

closed half-plane A half-plane together with its edge.

closed interval An interval on the real number line together with its endpoints. The closed interval from a to b is denoted $[a, b]$.

closed region A region that has no boundary or edge.

closed set A set that contains all of its boundary points.

closed surface A surface, such as that of a sphere, that has no boundary or edge.

closed trajectory A TRAJECTORY that forms a loop or closed curve.

closure grasp A secure grasp by a robot hand that does not allow any rotating or twisting of the object grasped.

clothoid *See* SPIRAL OF CORNU.

cluster point *See* LIMIT POINT.

C_n The CYCLIC GROUP of ORDER n.

co-punctal planes Three or more planes that have a point in common.

coaxal *See* COAXIAL.

coaxial In general, having the same axis.

coaxial circles Circles having the same RADICAL AXIS.

coaxial planes Planes that all contain a common line.

coaxial triangles Two triangles ABC and $A'B'C'$ are coaxial if the points of intersection of AB with $A'B'$, AC with $A'C'$, and BC with $B'C'$ are collinear.

cobordism theory The branch of ALGEBRAIC TOPOLOGY that studies the boundaries of manifolds.

cobweb *See* WEB DIAGRAM.

cochleoid The snail-like spiral that is the PERSPECTIVE PROJECTION of a circular helix. Its equation in polar form is $r = \dfrac{a \sin \theta}{\theta}$.

cochloid *See* CONCHOID OF NICOMEDES.

codeclination The complement of an angle of DECLINATION.

codomain The range of a function.

cofunction Each trigonometric function is paired with a cofunction: sine with cosine, tangent with cotangent, and secant with cosecant.

cofunction identities Identities that relate the trigonometric functions and their cofunctions.

cohomology group A GROUP belonging to a COHOMOLOGY THEORY.

cohomology theory An assignment of GROUPS, indexed by DIMENSION, to a topological space or MANIFOLD. A continuous function from a topological space S to another space T induces a HOMOMORPHISM from the cohomology groups of the space T to the cohomology groups of the space S with the same index.

collapsing compass A compass that can construct a circle with a given segment as its radius, but collapses when it is lifted from the paper.

collection *See* SET.

colline To be COLLINEAR, said of points or sets of points.

collinear Lying on a common line; said of points.

collineation A TRANSFORMATION that preserves collinearity of points and concurrency of lines.

collision In MOTION PLANNING, a collision is a coincidence of a robot with an obstacle.

color-preserving symmetry A SYMMETRY of a colored pattern that does not change the coloring.

color symmetry A SYMMETRY of a colored pattern that permutes, or rearranges, the colors.

colorable knot diagram A KNOT DIAGRAM that can be colored using three colors so that 1) each arc is colored by one color; 2) at least two colors are used; and 3) at each CROSSING, all three colors are used.

colored tiling A tiling in which each tile is assigned one of a finite number of colors. A colored tiling may or may not have the same SYMMETRIES as the underlying uncolored tiling.

column A vertical line of elements in a MATRIX.

column matrix A MATRIX with one COLUMN.

column vector *See* COLUMN MATRIX.

colunar triangles Two triangles on a sphere whose vertices are ANTIPODAL.

combinatorial geometry The study of the incidence properties of geometric objects. Combinatorial geometry includes the study of COVERINGS, PACKINGS, tilings, and coloring problems.

combinatorial manifold *See* TRIANGULATED MANIFOLD.

combinatorial structure For a polyhedron, graph, or map, the number and arrangement of vertices, edges, and faces. For example, a polyhedron and its SCHLEGEL DIAGRAM have the same combinatorial structure.

combinatorial symmetry A permutation of the vertices of a polyhedron or graph that preserves adjacency. A combinatorial symmetry is not necessarily an ISOMETRY.

combinatorial topology The topology of SIMPLICES, SIMPLICIAL COMPLEXES, and topological spaces that are topologically equivalent to simplicial complexes.

commensurable Having a common measure. For example, two segments are commensurable if there is another segment whose measure goes evenly (without remainder) into the measure of the two segments. Segments with lengths 1 and 1/8 are commensurable, but segments with lengths 1 and $\sqrt{2}$ are not.

common Shared. For example, points lie on a common line if they are collinear.

common cycloid *See* CYCLOID.

common external tangent A line tangent to two circles that does not go between the two circles; it does not intersect the segment containing their centers.

common internal tangent A line tangent to two circles that goes between them; it intersects the segment containing their centers.

common notion An AXIOM, especially one that applies to all branches of mathematics.

common tangent line A line that is TANGENT to two or more circles or curves.

common tangent plane A plane that is TANGENT to two or more spheres or surfaces.

commutative Describing a binary operation having the property that changing the order of operation results in the same answer. For example, addition and multiplication are commutative.

compact set A CLOSED, BOUNDED set.

compass An instrument used to construct a circle.

compass dimension *See* DIVIDER DIMENSION.

compatible coloring *See* PERFECT COLORING.

compatible transformation A TRANSFORMATION T is compatible with a transformation S if the composition TS exists; this will happen if and only if the IMAGE of S is contained in the DOMAIN of T.

complement of a set *See* COMPLEMENTARY SETS.

complement of an angle An angle whose measure added to the measure of the given angle gives 90°.

complementary angles Two angles whose measures add up to 90°.

complementary sets Two disjoint subsets of a given set are complementary if their UNION is the given set. Each set is the complement of the other with respect to the given set.

complete An AXIOMATIC SYSTEM is complete if any statement involving concepts based on the axioms can be either proved or disproved within the system.

complete bipartite graph A BIPARTITE GRAPH in which each vertex in one set of vertices is connected to every vertex in the complementary set of vertices. It is denoted $K_{m, n}$, where there are m vertices in one set and n vertices in the other.

complete graph A GRAPH in which each pair of vertices is connected by one edge.

complete pencil The PENCIL of all lines through a given point.

complete quadrangle A set of four coplanar points, no three of which are collinear, together with the six lines joining these points in pairs.

complete quadrilateral A set of four coplanar lines, no three concurrent, together with the six vertices that are the points of intersection of pairs of these lines.

complete range A RANGE of points consisting of all the points on a line.

complete triangle A set of three noncollinear points together with the three lines connecting these points in pairs.

complex analysis Techniques from CALCULUS applied to functions from the complex numbers to the complex numbers.

complex conjugate The complex conjugate of the complex number $a + bi$ is the complex number $a - bi$. The product of a complex number and its complex conjugate is a real number.

complex manifold A space in which every point is contained in a NEIGHBORHOOD that looks like the COMPLEX PLANE or a Cartesian product of complex planes.

complex number　A number of the form $a + bi$ where a and b are real numbers and $i^2 = -1$.

complex plane　The Cartesian plane in which each point corresponds to a complex number. The x-axis corresponds to the real numbers and the y-axis corresponds to PURELY IMAGINARY NUMBERS. The point (a, b) represents the complex number $a + bi$.

complex vector space　A VECTOR SPACE whose field of scalars is the complex numbers.

component　(1) A part of a shape, space, or graph that is CONNECTED. It is also called a connected component. (2) The scalar coefficient of one of the VECTORS in the VECTOR RESOLUTION of a given vector.

compose　To apply two or more functions in sequence.

composite function　*See* COMPOSITION.

composite knot　A KNOT that is the CONNECTED SUM of two nontrivial knots.

composite number　A positive integer that can be factored as the product of two or more PRIMES. Thus, 4, 6, 8, and 9 are composite numbers.

composition　The function resulting from applying two or more functions in sequence.

composition of a tiling　A new tiling whose tiles are the unions of adjacent tiles in another tiling.

compound curve　A connected curve made up of parts of circles that are tangent at their points of intersection.

compound polygon　A polygon determined by a set of congruent REGULAR POLYGONS with a common center. For example, the Star of David is a compound polygon formed from two equilateral triangles.

compound polyhedron　A polyhedron that is the union of a set of congruent REGULAR POLYHEDRA with a common center. For example, the stellated octahedron is a compound polyhedron formed from two tetrahedra.

compression　A TRANSFORMATION of the plane given by the equations $f(x, y) = (x, ky)$ or $f(x, y) = (kx, y)$ where $0 < k < 1$.

computational geometry　The study of algorithms or procedures used for solving geometric problems on a computer. Computational geometry includes the design of algorithms as well as the analysis of their effectiveness and efficiency.

computer animation The use of a computer and computer graphics to make animated cartoons or films.

computer graphics The science of generating, storing, manipulating, and displaying artwork with the aid of a computer.

computer vision The use of a computer to analyze and interpret visual data.

concave Not CONVEX.

concentric circles Two or more circles with the same center that are located in the same plane.

concho spiral A helix lying on a cone with the property that the tangents to the concho spiral intersect the GENERATORS of the cone at a constant angle.

conchoid A curve determined from a fixed curve C, a fixed point O, and a fixed distance k. The two points at a distance k from the intersection of a line through the point O with the curve C are on the conchoid.

conchoid of Nicomedes A curve that can be used to trisect angles and duplicate the cube. It is the conchoid determined by a line and a point not on the line and is given by the equation $(x^2 + y^2)(y - a)^2 = k^2 y^2$, where a and k are constants.

conclusion (1) The statement at the end of a proof that tells what has been proved. (2) The statement q in the implication "If p, then q."

concur To intersect at exactly one point; said of lines.

concurrent lines Lines that intersect at exactly one point.

concyclic points Points that lie on the same circle.

condensation point For a given set, a point with the property that any of its NEIGHBORHOODS contains an uncountable infinity of points from the set.

conditional *See* IMPLICATION.

cone (1) The surface consisting of segments connecting points on a closed planar curve, called the base, to a noncoplanar point, called the vertex of the cone. (2) The surface swept out by a ray or line that is rotated about an axis passing through the endpoint of the ray or a point on the line.

cone angle The angle at the vertex of a cone. It is measured by cutting the cone along a GENERATOR of the cone, flattening the cone, and measuring the resulting planar angle between the edges of the flattened cone.

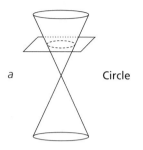

a Circle

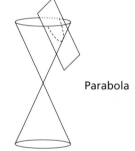

b Ellipse

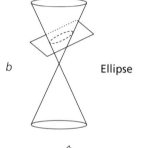

c Parabola

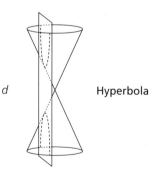

d Hyperbola

The four conic sections

cone point The vertex of a cone.

configuration A finite collection of *p* points and *l* lines or segments such that each point is incident with λ lines and each line is incident with π points. In any configuration, $p\lambda = l\pi$. A configuration may also consist of points and planes, each point incident to the same number of planes, and each plane containing the same number of points.

confocal Having the same focus or foci, said of a family of curves or surfaces.

conformal mapping A function that preserves angle measure between any two lines.

conformal projection A PROJECTION that preserves the angle measure between any two lines.

congruence *See* ISOMETRY.

congruent Having equal measurements, therefore having the same size and shape.

congruent polygons Two polygons for which there is a one-to-one correspondence between angles and sides with corresponding angles congruent and corresponding sides congruent.

congruent transformation *See* ISOMETRY.

conic *See* CONIC SECTION.

conic section A conic section is a curve formed by the intersection of a plane with a cone. There are four types of conic sections depending on how the plane meets the cone: circle, ellipse, parabola, and hyperbola. The circle results if the plane is perpendicular to the axis of the cone. The ellipse results if the plane cuts off one NAPPE of the cone in a closed curve; the circle is thus a special case of the ellipse. The parabola results if the plane is parallel to a line on the cone. The hyperbola results if the plane intersects both nappes of a double cone.

The graph of a second-degree polynomial equation in *x* and *y* is a conic section. The equation $ax^2 + by^2 + cxy + dx + ey + f = 0$ gives a circle if *a* and *b* are equal, an ellipse if *a* and *b* have the same sign, a hyperbola if *a* and *b* have opposite signs, and a parabola if either *a* or *b* is zero.

The conic sections can also be defined in terms of their relationships to specific points (foci) or lines (directrices). The circle is the set of all points at an equal distance from a fixed point, its center. A point on an ellipse has the property that its distances from two fixed points (its foci) have a constant sum. A point on a parabola has the property that it distance from a fixed point (the focus) is equal to its distance from a fixed line (the directrix). A point on a hyperbola has the property that its distances from two fixed points (the foci) have a constant difference. Further, the conics can be defined in terms of the ratio (the eccentricity) of the distance from a point on the conic to a fixed point (the focus) and the distance from the point to a line (the directrix). An ellipse has eccentricity less than 1; a parabola has eccentricity 1; and a hyperbola has eccentricity greater than 1.

conicoid *See* QUADRIC.

conjugate A term used to indicate that two objects have a special correspondence with each other; the nature of the correspondence depends on the context.

conjugate angles Two angles whose measures add up to 360°.

conjugate arcs Two arcs whose union is a circle and which intersect only at their endpoints.

conjugate axis A segment perpendicular to the TRANSVERSE AXIS of a hyperbola. The ASYMPTOTES of a hyperbola are the diagonals of the rectangle determined by the transverse axis and the conjugate axis.

conjugate lines (1) Two lines are conjugate if each contains the image of the other under a POLARITY. (2) Two lines are conjugate if each contains the POLE, with respect to a given circle of INVERSION, of the other.

conjugate points (1) Two points are conjugate if each contains the POLAR, with respect to a given circle of INVERSION, of the other. (2) Two points are conjugate if each lies on the image of the other under a POLARITY.

conjugate triangles Two triangles such that each vertex of one triangle is the POLE with respect to a given circle of a side of the other.

conjunction A statement formed by joining two statements with the word *and.*

connected sum A new KNOT formed from two given knots by making a cut in each knot and then gluing each of the two new ends of one knot to a new end of the other knot.

connectivity For a GRAPH, the number of edges that must be removed to produce a tree.

conoid The surface formed by revolving a parabola or one sheet of a hyperbola about its axis of symmetry.

consecutive Adjacent or following one another in order. For example, 3 and 4 are consecutive integers.

consecutive angles Angles of a polygon that share a side.

consecutive vertices Vertices of a polygon that are the endpoints of the same side.

consequent The second term of a ratio. The consequent of the ratio *a:b* is *b*.

conservative field A vector field that is the GRADIENT FIELD of a SCALAR FIELD.

consistency The property of a collection of AXIOMS that contradictory theorems cannot be proved from the axioms.

constant Not changing, having a fixed value. Constant is the term used to describe a variable that represents a number with a fixed but unknown, rather than varying, value. A constant function has the same value for every point in the domain.

constant curvature, surface of A surface with constant nonzero GAUSSIAN CURVATURE.

constant of inversion The square of the RADIUS OF INVERSION.

constant of proportionality *See* DILATION.

constrained triangulation A TRIANGULATION that must include certain specified edges.

constructible number A segment having length equal to a constructible number can be constructed with compass and straightedge.

construction The drawing of a figure using compass and straightedge.

contact number *See* KISSING NUMBER.

content A measure, such as area or volume, of the space enclosed by a closed curve or surface.

contingency, angle of The angle between two tangent vectors at nearby points on a curve.

continued proportion An equality of two or more ratios. The equality 1/2 = 2/4 = 4/8 is written as the continued proportion 1:2:4:8.

continuity, axiom of The statement that every CONVERGENT SEQUENCE of points in a geometry has a limit point contained in the geometry. The set of real numbers satisfies the axiom of continuity, but the set of rational numbers does not since the sequence 1, 1.4, 1.41, 1.414, . . . converges to $\sqrt{2}$, which is not rational.

continuous function A function that preserves the nearness of points. The graph of a continuous function from the real numbers to the real numbers does not have any breaks or gaps. A continuous function of topological spaces has the property that the inverse image of an OPEN SET in the RANGE is an open set of the DOMAIN.

continuum The real number line.

Continuum Hypothesis The statement that the cardinality of the continuum of real numbers is the next largest cardinality after the cardinality of the natural numbers. It is independent of the usual axioms of set theory.

contour line The PROJECTION onto a plane of the intersection of a surface with a plane parallel to the projection plane.

contractible Describing a set in a topological space that can be DEFORMED to a point.

contraction (1) A DILATION for which the SIMILARITY RATIO has absolute value less than 1. (2) A SIMILARITY for which the RATIO OF SIMILITUDE is less than 1. (3) A function f such that the distance between $f(x)$ and $f(y)$ is always less than or equal to a multiple s of the distance between x and y. The number s is called the contractivity factor of the contraction.

contractive mapping *See* CONTRACTION.

contradiction A statement that is false.

contrapositive The contrapositive of the implication $p \to q$ is the implication $\sim q \to \sim p$. The contrapositive of an implication has the same truth value as the implication.

converge To get closer and closer together. Two curves converge if there are points, one on each curve, that are closer together than any given distance, however small. A sequence of numbers $a_1, a_2, a_3, \ldots$ converges to a number L, called the limit of the sequence, if the difference between a_n and L can be made smaller than any given number, no matter how small, when n is large enough.

convergent sequence A sequence of points or numbers that CONVERGES to a limit.

converse The converse of the implication $p \to q$ is the implication $q \to p$.

converse form of a definition An inverted form of a definition. For example, the converse form of the definition "A triangle is a polygon with three sides" is the statement "A polygon with three sides is a triangle." A definition and its converse are equivalent.

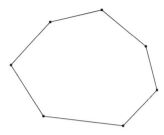

Convex polygon

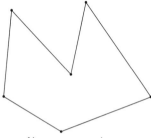

Nonconvex polygon

Convex

convex A figure is convex if a line segment connecting any two points in the interior of the figure lies entirely within the interior of the figure. A polygon is convex if each interior angle has measure less than 180° and a polyhedron is convex if the sum of the angle measures at each vertex is less than 360°. The empty set and a set containing one point are convex by convention.

convex body A CONVEX set that is CLOSED, BOUNDED, NONEMPTY, and has at least one interior point.

convex combination A sum of SCALAR MULTIPLES of one or more vectors where the scalars are non-negative and their sum is 1. A convex combination of two or more vectors lies in their CONVEX HULL.

convex cover *See* CONVEX HULL.

convex geometry The study of CONVEX subsets of Euclidean space and their properties.

convex hull The smallest CONVEX set containing a given figure or set of points.

Conway polynomial A polynomial determined from the sequence of CROSSINGS in a KNOT or LINK. It is a KNOT INVARIANT.

Conway worm A long zigzag strip in a PENROSE TILING that looks like a worm.

Conway's criterion Two conditions, either of which guarantees that a six-sided PROTOTILE can tile the plane. The first condition says that each side of the tile is a TRANSLATION of the opposite side. The second condition says that a pair of opposite sides are translates of each other and that each of the other four sides has POINT SYMMETRY.

coordinate *See* COORDINATE PLANE.

coordinate axis A copy of the real number line used to determine coordinates of a point in a coordinate system.

coordinate net A set of local coordinates used for a small region on a surface.

coordinate plane A plane in which each point is assigned numbers, called coordinates, that determine its exact location in the plane. The real coordinate plane has two intersecting copies of the real number line, called the coordinate axes. The point of intersection of the axes is called the origin. The coordinates of a point on the plane are the real numbers corresponding to the intersection of lines through the given point parallel to the two axes. Coordinates (a, b) determine the point at the intersection of the line parallel to the *y*-axis through *a* on the *x*-axis and the line parallel to the *x*-axis through *b* on the *y*-axis.

coordinate space A space in which each point is assigned coordinates that determine its point in the space. A coordinate space is the generalization of COORDINATE PLANE to any dimension.

coordination number *See* KISSING NUMBER.

coplanar Lying on a common plane, said of points, lines, or planar figures.

coplane To be COPLANAR.

copolar points Points having a common POLAR.

copolar triangles Two triangles *ABC* and *A'B'C'* are copolar if the three lines *AA'*, *BB'*, and *CC'* are concurrent.

core The largest CONVEX solid that can be constructed inside a given polyhedron.

corner A vertex of a polygon or a point on a curve where there is an abrupt change of direction.

cornicular angle The angle between an arc of a circle and a tangent to the circle; the vertex of a cornicular angle is the point of tangency.

corollary A THEOREM that has been derived directly from another theorem.

correlation In PROJECTIVE GEOMETRY, a correspondence between points and lines that preserves INCIDENCE.

corresponding angles (1) For a pair of similar polygons, corresponding angles are a pair of congruent angles, one in each polygon, with the same relative positions in the polygons. (2) For a transversal cutting parallel lines, corresponding angles are translations of each other, on the same side of the transversal.

corresponding points Two points, one on each of two lines, such that the two angles formed on the same side of the segment connecting the two points are congruent.

corresponding sides For a pair of similar polygons, sides opposite corresponding angles.

cos *See* COSINE.

cos⁻¹ *See* INVERSE COSINE.

cosecant For an acute angle in a right triangle, the ratio of the hypotenuse to the opposite side. A directed angle with vertex at the origin of the Cartesian plane determines a point (x, y) on the unit circle centered at the origin; the cosecant is $1/y$.

cosh   *See* HYPERBOLIC COSINE.

cosine   For an acute angle in a right triangle, the ratio of the adjacent side to the hypotenuse. A directed angle with vertex at the origin of the Cartesian plane determines a point (x, y) on the unit circle centered at the origin; the cosine is x.

cosine circle   *See* SECOND LEMOINE CIRCLE.

cot   *See* COTANGENT.

cot^{-1}   *See* INVERSE COTANGENT.

cotangent   For an acute angle in a right triangle, the ratio of the adjacent side to the opposite side. A directed angle with vertex at the origin of the Cartesian plane determines a point (x, y) on the unit circle centered at the origin; the cotangent is x/y.

counterexample   A counterexample is a particular object that does not have a particular characteristic under consideration; it demonstrates the purpose or value of the characteristic. For example, skew lines do not intersect and are not parallel, showing that the requirement that lines be COPLANAR is necessary in the definition of parallel lines.

course line   The planned path of a ship or plane.

covering   A collection of subsets of a region whose union is the whole region.

covering space   A space that maps onto another space, called the base space, with the property that it is a HOMEOMORPHISM near any point in the domain. For example, the real number line is a covering of the unit circle in the complex plane using the mapping $x \mapsto e^{ix}$.

coversed sine   *See* VERSED SINE.

coversine   *See* VERSED SINE.

critical point   A point on a curve or surface where the tangent line or plane is parallel to the x-axis or the xy-plane, respectively.

critical value   The value a function takes on at a critical point.

cross   A set of n mutually orthogonal segments with a common midpoint.

cross joins   The lines AB' and $A'B$ are the cross joins of the ordered pair of points A and A' and the ordered pair of points B and B'.

cross multiply   *See* RULE OF THREE.

cross polytope   A POLYTOPE whose vertices are the endpoints of the segments of a CROSS.

cross product The cross product of two VECTORS **v** and **w** is the vector, denoted **v** × **w**, that is perpendicular to the plane determined by **v** and **w**, has length equal to the area of the parallelogram formed by **v** and **w**, and has direction so that **v**, **w**, and **v** × **b** form a right-handed system. If **v** or **w** have the same or opposite directions, or if one of them is the zero vector, then their cross product is the zero vector. The cross product is an ANTICOMMUTATIVE operation.

cross ratio For four collinear points, A, B, C, and D, the cross ratio (AB, CD) is $\dfrac{AC/CB}{AD/DB}$ where directed distances between the points are used. The cross ratio of a harmonic set is -1.

cross ratio of a cyclic range The cross ratio of four points on a circle is the CROSS RATIO of a PENCIL of lines connecting any point on the circle, distinct from the four points, to the four points.

cross ratio of a pencil The cross ratio of a pencil of four lines is the CROSS RATIO of the four points of intersection of the pencil with any TRANSVERSAL that does not pass through the vertex of the pencil.

cross section *See* SECTION.

crosscap A MÖBIUS BAND attached to a surface by attaching or gluing each point on the boundary of the Möbius band to a point on boundary of a circular hole in the surface. This can be done since the boundary of the Möbius band is a circle; however, it can be done without self-intersections only in four-dimensional space.

crossed hyperbolic rotation An EQUIAFFINITY of the plane that exchanges the branches of a given hyperbola.

crossed quadrilateral A CONVEX quadrilateral.

crossing (1) A point that is not a vertex of a graph but is a point of intersection of two edges in a planar representation of the graph. (2) *See* CROSSING POINT.

crossing index The crossing number of a knot. *See* CROSSING NUMBER.

crossing number (1) The number of times a segment crosses the boundary of a region. (2) The smallest possible number of crossings in a diagram of a given knot. It is a KNOT INVARIANT.

crossing point A break in an arc in a KNOT DIAGRAM that allows another arc to pass through.

crunode *See* NODE.

crystal A regular arrangement of atoms or molecules.

crystal class A category of crystals based on properties of the point groups. There are 32 different crystal classes corresponding to the 32 different CRYSTALLOGRAPHIC POINT GROUPS.

crystal system A category of crystals based on properties of the 32 CRYSTALLOGRAPHIC POINT GROUPS. There are seven crystal systems: triclinic, monoclinic, orthorhombic, rhombohedral, tetragonal, hexagonal, and cubic. Each crystal system contains two or more crystal classes.

crystallographic group The SYMMETRY GROUP of a three-dimensional LATTICE. There are 230 different crystallographic groups and every crystal has one of these groups as its symmetry group.

crystallographic point group The symmetry group of a three-dimensional lattice that leaves a specific point fixed. There are 32 different crystallographic point groups.

crystallographic restriction The restriction that a ROTATIONAL SYMMETRY of a tiling or LATTICE can have no order except 2, 3, 4, or 6.

crystallographic space group *See* CRYSTALLOGRAPHIC GROUP.

csc *See* COSECANT.

csc⁻¹ *See* INVERSE COSECANT.

cube (1) A polyhedron having six congruent faces, each of which is a square. (2) The third power of a number or expression.

cubic Related to a cube, either the polyhedron or the third power of an expression.

cubic close packing A PACKING of three-dimensional space by spheres in which the centers of the spheres are at the points of a LATTICE formed of CUBOCTAHEDRA. Each sphere is tangent to 12 other spheres.

cubic lattice A LATTICE whose LATTICE UNIT is a cube. It is also called a cubical lattice.

cubic packing A PACKING by spheres of three-dimensional space in which the centers of the spheres form a CUBIC LATTICE. Each sphere touches six other spheres.

cubic parabola A curve that is the graph of a cubic equation such as $y = ax^3 + bx^2 + cx + d$, where a, b, c, and d are constants.

cubic spline A collection of arcs of cubic curves that make up a smooth curve containing a given set of points. Cubic splines are used by computer graphics programs to draw a smooth curve through a given set of points.

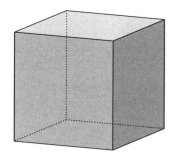

Cube

cuboctahedron An SEMIREGULAR POLYHEDRON with eight equilateral triangular faces and six square faces.

cubohemioctahedron A NONCONVEX polyhedron with six square faces and four hexagonal faces. It is a faceted cuboctahedron.

cuboid *See* RECTANGULAR PARALLELEPIPED.

curl A vector field that gives a measure of the circulation of a given vector field per unit area.

curtate cycloid *See* CYCLOID.

curvature A measure of the degree of turning or winding of a curve at a point. It is obtained from the AVERAGE CURVATURE over very small intervals containing the point and is 0 for a straight line.

curvature centroid The CENTER OF GRAVITY of an arc weighted with mass whose density at a point is proportional to the curvature at that point.

curvature vector For a curve or surface, a VECTOR perpendicular to a tangent line or plane with magnitude equal to the CURVATURE at that point.

curve A one-dimensional shape.

curve fitting The process of finding the equation of a curve that passes through a given set of points.

curve of constant slope *See* HELIX.

curve of constant width A CLOSED curve with the property that the width enclosed by the curve is the same in every direction.

curve of double curvature A nonplanar curve in three-dimensional space.

curved line *See* CURVE.

curves of the second order The conic sections; so called because the equations for their graphs are second-order equations in x and y.

curvilinear Having sides, edges, or faces that are not RECTILINEAR.

curvilinear coordinates Coordinates for a point on a surface with respect to a NET OF CURVES on the surface.

cusp (1) A point on a curve where two different arcs of the curve meet but where the tangent line is still well defined because both branches have the same tangent. (2) A singularity of a function from a surface to a surface that occurs when the domain surface gets "pleated" before it is mapped to the range. The cusp point has one inverse image; each point in the interior of the pleat has three inverse images; each point on the edge of the cusp has two inverse images; and each point outside the cusp has one inverse image.

cusp of the first kind A cusp where there is a branch of the curve on each side of the tangent line at the cusp.

cusp of the second kind A cusp where both branches of the curve are on the same side of the tangent line at the cusp.

cusp point *See* CUSP.

cut line A curve on a CONNECTED surface with the property that when it is removed the surface is no longer connected.

cut point A point on a CONNECTED curve with the property that when it is removed the curve is no longer connected. For example, the endpoint of a segment is not a cut point of the segment but any interior point is a cut point.

cycle (1) A permutation of elements $a_1, a_2, a_3, \ldots, a_n$ that takes a_1 to a_2, a_2 to $a_3, \ldots$, and a_n to a_1. (2) For a given function f, a finite set of points $x_1, x_2, \ldots, x_n$ such that $f(x_1) = x_2, f(x_2) = x_3, \ldots$, and $f(x_n) = x_1$.

cyclic (1) Describing a polygon that can be inscribed in a circle. (2) Describing a design whose SYMMETRIES are only ROTATIONS or a describing a SYMMETRY GROUP that contains only rotations.

Cycloid

cyclic group The SYMMETRY GROUP of a rosette or circle; it consists only of rotations.

cyclic range A set of points all lying on the same circle.

cyclides The surfaces that make up the ENVELOPE of a family of spheres tangent to three fixed spheres. The lines of curvature of a cyclide are circles. The TORUS is an example of a cyclide. Also called the cyclides of DUPIN.

cycloid The locus traced out by a point attached to a circle as the circle rolls along a line without slipping. A common cycloid is generated by a point on the circumference of the circle, an oblate cycloid is generated by a point outside the circle, and a curtate cycloid is generated by a point inside the circle.

cycloidal pendulum A pendulum with a bob on a string that is restricted to oscillate between two fixed arcs of a CYCLOID. The bob of a cycloidal pendulum traces out an arc of a cycloid. It was invented by Dutch physicist Christian Huygens.

cyclotomic Dividing a circle into congruent arcs.

cylinder A solid with two congruent bases connected by a lateral surface generated by segments connecting corresponding points on the two bases.

Cylinder

cylindrical coordinates Coordinates for space with respect to two fixed axes meeting at the origin. The coordinates (r, θ, z) correspond to a point in space where r is the distance of the point along the horizontal axis and θ gives the directed angle from a fixed horizontal axis to the ray from the origin to the point and z gives the distance along the vertical axis.

cylindrical group The SYMMETRY GROUP of a pattern on the surface of a cylinder.

cylindrical helix A helix on the surface of a cylinder.

cylindrical projection PROJECTION of a sphere onto a cylinder that is tangent to the sphere.

cylindrical symmetry *See* RADIAL SYMMETRY.

cylindroid A surface consisting of all lines that intersect each of two curves and that are parallel to a given line.

dale A region on a TERRAIN whose descending SLOPE LINES run to the same PIT.

Dandelin sphere A sphere tangent to a circular cone and to a plane intersecting the cone. A DANDELIN sphere is tangent to the plane at the directrix of the conic section determined by the plane.

Darboux vector For a point on a curve in three-dimensional space, the vector $k\mathbf{b} + \tau\mathbf{t}$, where k is the curvature, $\mathbf{b}$ is the binormal vector, τ is the torsion and $\mathbf{t}$ is a unit tangent vector.

dart One of the two PENROSE TILES. It is a DELTOID.

dead reckoning Determination of a ship's position from compass readings, speed, and time, rather than from astronomical observations.

dead wind *See* HEAD WIND.

decagon A polygon with 10 sides.

decagram The star polygon created from the decagon.

decahedron A polyhedron with 10 faces.

decapod A region that is the union of 10 congruent isosceles triangles with one vertex in common. The ratio of leg to base for each triangle is $\tau/(\tau + 1)$. There are 62 possible decapods.

declination (1) The angular distance between the celestial equator and the object of observation. A northern declination is positive and a southern declination is negative. (2) The difference between magnetic north, determined by a compass, and true north.

decomposition of a tiling Creation of a new tiling whose tiles are formed from a given tiling by cutting up the tiles of the original tiling.

decomposition of a vector The expression of a vector as a linear combination of finitely many vectors, usually the basis vectors of a vector space.

Dedekind cut A Dedekind cut gives a precise mathematical description of a real number using the set of rationals. A Dedekind cut consists of two disjoint sets of rationals whose union contains all the rationals; every element of one set is less than every element of the other set and both sets have no largest element. If the set of larger numbers has a least element, that number is the rational real number described by the cut. If it does not have a least element, the cut describes the irrational number between the two sets. For example, the cut that describes $\sqrt{2}$ consists of the set of rationals whose square is greater than 2 and the set of rationals whose square is less than 2.

deduction The process of deriving a specific conclusion from a general principle.

defect *See* ANGULAR DEFECT.

deferent In the construction of an EPICYCLOID, the fixed circle on which the moving circle rolls.

deficiency *See* ANGULAR DEFICIENCY.

defined term An object in a mathematical theory that is defined using the UNDEFINED TERMS of the theory. *See* AXIOMATIC SYSTEM.

definition A statement giving the mathematical properties of a term using UNDEFINED TERMS and terms that have already been defined. *See* AXIOMATIC SYSTEM.

deflection, angle of The angle between a ray of light striking an object, such as a raindrop, and the ray of light leaving the object.

deform To transform by a DEFORMATION.

deformation A continuous sequence of transformations that involve only stretching or shrinking. For example, a circle in three-dimensional space can be deformed into a square but not into a trefoil knot.

degenerate Satisfying a definition in a minimal way so that the result is of little interest. For example, three collinear points determine a degenerate triangle.

degree   A unit of angle measure; one 360th of a circle.

degree of a polynomial   The degree of a monomial is the sum of the powers of the variables. For example, x^5 and x^3y^2 both have degree 5. The degree of a polynomial is the degree of the term with highest degree.

degree of a vertex   In a polyhedron, the degree of a vertex is the number of edges incident to the vertex. In a graph, the degree of a vertex is the number of edges incident to the vertex.

degree of freedom   The number of parameters or coordinates necessary to describe a geometric structure. For example, a circle has one degree of freedom and the surface of a sphere has two degrees of freedom.

Dehn surgery   In DEHN surgery, a tubular NEIGHBORHOOD about a KNOT in a three-dimensional sphere is removed and a solid TORUS is glued in along the boundary using information from a TORUS KNOT on the torus.

del   *See* NABLA.

Delaunay triangulation   A TRIANGULATION with the property that the CIRCUMCIRCLE of a triangle in the triangulation contains no vertices of the triangulation other than those of the triangle itself.

Delian problem   *See* DUPLICATION OF THE CUBE.

deltahedron   A CONVEX polyhedron whose faces are triangles. There are eight deltahedra whose faces are equilateral triangles, having four, six, eight, 10, 12, 14, 16, or 20 faces.

deltoid   (1) A CONCAVE quadrilateral with two pairs of adjacent sides congruent. (2) A HYPOCYCLOID with three cusps.

deltoidal hexecontahedron   *See* TRAPEZOIDAL HEXECONTAHEDRON.

deltoidal icositetrahedron   *See* TRAPEZOIDAL ICOSITETRAHEDRON.

dendrite   A fractal shaped like a tree.

dense subset   A subset of a topological space whose INTERIOR and BOUNDARY make up the whole space.

density   (1) For a given solid, the ratio of mass to volume. (2) For a STAR POLYGON, the number of edges intersected by a ray from the center of the star polygon that does not intersect any of the vertices. For example, the pentagram has density 2. (3) A measure of the efficiency of a PACKING. For a two-dimensional packing, it is the ratio of the area of the sets in the packing to the total area of the region. For a three-dimensional packing, it is the ratio of the volume of the sets in the packing to the total volume of the region.

denying the premise A logical fallacy in which one tries to conclude the negation of a statement q from the two statements $p \to q$ and the negation of p.

dependent variable A variable whose value depends on the values of other variables.

derivative In CALCULUS, the measure of the rate of change of the value of a function with respect to its independent variable.

Desargues, configuration of A configuration with 10 points and 10 lines such that each line contains three points and each point lies on three lines.

Desargues, finite geometry of A FINITE GEOMETRY with exactly 10 points and 10 lines. Each line contains three points and each point lies on three lines.

Desarguesian space A projective space in which the theorem of Desargues is valid.

Descartes formula The DESCARTES formula states that the total angular deficiency of a polyhedron is 720°.

descending slope line *See* SLOPE LINE.

descriptive geometry PROJECTIVE GEOMETRY, especially that part used in PERSPECTIVE.

det *See* DETERMINANT.

determinant A number computed from all the elements in a square matrix that gives valuable information about the matrix and the LINEAR TRANSFORMATION represented by the matrix. For example, if the determinant of a matrix is 0, the matrix does not have a multiplicative inverse. If the sign of the determinant is positive, the linear transformation preserves orientation, and if the sign is negative, it reverses orientation. The determinant of the 2×2 matrix $\begin{bmatrix} a & b \\ c & d \end{bmatrix}$ is $ad - bc$.

determinant of a lattice The volume of a UNIT CELL of a LATTICE.

determinant straight line *See* SEGMENT.

deterministic Predictable; governed by a mathematical rule or formula.

deterministic dynamical system A DYNAMICAL SYSTEM whose behavior is described exactly by mathematical formulation.

deuce One of the seven different VERTEX NEIGHBORHOODS that can occur in a PENROSE TILING.

developable surface A surface whose GAUSSIAN CURVATURE is 0 everywhere. For example, a cylinder and a cone are developable surfaces.

dextro An ENANTIOMER that rotates polarized light to the right.

diacaustic *See* CAUSTIC CURVE.

diagonal A segment connecting two nonadjacent vertices of a polygon.

diagonal 3-line *See* DIAGONAL TRILATERAL.

diagonal 3-point *See* DIAGONAL TRIANGLE.

diagonal matrix A square matrix whose elements that are not on the MAIN DIAGONAL are 0.

diagonal of a circuit An edge that connects two vertices of a CIRCUIT in a graph but is not part of the circuit.

diagonal point The point of intersection of two opposite sides of a COMPLETE QUADRANGLE.

diagonal triangle The triangle determined by the three diagonal points of a COMPLETE QUADRANGLE. If the three points are collinear, the triangle will be degenerate.

diagonal trilateral The figure formed by the three lines joining opposite vertices of a COMPLETE QUADRILATERAL.

diagram (1) A rough sketch of a geometric figure, not necessarily drawn to scale. (2) A symbolic representation of a geometry as a graph. One node represents the points of the geometry, another the lines, and so on. A labeled edge connecting two nodes represents the geometric relationships between the objects represented by the nodes.

diameter A chord that passes through the center of a circle.

diameter of a set The length of the longest segment contained in the set.

diamond A RHOMBUS whose acute angles are 45°. Sometimes, a diamond is a rhombus that is not a square.

Dido's problem The problem of finding the maximum area that can be surrounded by a curve of fixed length together with a segment of any length. This problem arose when Queen Dido was given as much land along the seashore as she could surround by an ox hide; she cut the hide into thin strips, tied them together, and got enough land to found the city of Carthage.

diffeomorphism A HOMEOMORPHISM that is DIFFERENTIABLE.

difference equation An equation that gives the value of a number in a sequence as a function of one or more preceding numbers in the sequence.

differentiable Describing a curve or surface having a tangent line or tangent plane at every point. A circle is differentiable but a square is not.

differentiable function A function that has a well-defined DERIVATIVE at each point in its domain.

differentiable manifold A MANIFOLD that has a well-defined tangent space at each point.

differential equation An equation that uses DERIVATIVES to describe how changing quantities are related to one another.

differential geometry The study of curves and surfaces using the tools of CALCULUS.

differential geometry in the large That part of DIFFERENTIAL GEOMETRY focused on the study of the behavior of a curve or surface as a whole.

differential geometry in the small That part of DIFFERENTIAL GEOMETRY focused on the study of the local behavior of curves and surfaces.

differential topology The study of MANIFOLDS using techniques from CALCULUS.

digon A polygon with two vertices and two sides. In the plane, the two sides coincide, but on a sphere, a digon is wedge-shaped.

digraph A GRAPH in which each edge has been assigned a direction from one of its vertices to the other.

dihedral Describing a design whose SYMMETRIES include REFLECTIONS or describing a SYMMETRY GROUP that contains reflections.

dihedral angle The union of two half-planes that share an edge, together with the shared edge. The measure of a dihedral angle is equal to the measure of an angle formed by two intersecting rays, one on each half-plane, that are perpendicular to the shared edge.

dihedral content For a polyhedron, the sum over all edges of the TENSORS $l \otimes a$ where l is the length of the edge and a is the measure of its dihedral angle.

dihedral group A group of SYMMETRIES that contains only ROTATIONS and REFLECTIONS. It is the symmetry group of a dihedral design.

dihedral tiling A tiling in which each tile is congruent to one of two distinct PROTOTILES.

dihedron A map on a sphere consisting of two regions.

Dijkstra's algorithm An algorithm for finding the shortest path in a
WEIGHTED GRAPH.

dilatation *See* DILATION. In some contexts, a dilatation is a transformation
with the property that a line and its image are parallel, so it is either a
dilation, a TRANSLATION, or an INVERSION.

dilatative rotation *See* DILATIVE ROTATION.

dilation A DIRECT SIMILARITY of the plane having a fixed point, called the
center of dilation. A dilation multiplies every distance by a constant
ratio $|r|$, where r is a nonzero number called the ratio of dilation.
Under a dilation a point and its image are collinear with the center of
dilation C; a point and its image are on the same side of C if the ratio
of dilation is positive and on different sides of C if the ratio of
dilation is negative

dilative homography The product of two INVERSIONS with respect to
nonintersecting circles.

dilative reflection The composite of a REFLECTION with a DILATION whose
center is on the line of reflection.

dilative rotation The composition of a CENTRAL DILATION with a ROTATION
about the same center.

dimension (1) A measurement of an object; for example, the dimensions of a
rectangle are its length and width. (2) The number of degrees of
freedom in a geometric space. The dimension of a point is 0, of a line
or curve, it is 1, of a plane or surface, it is 2, and so on. (3) For a
VECTOR SPACE, the dimension is the number of vectors in a BASIS.

dimorphic tile A PROTOTILE that admits exactly two different MONOHEDRAL
TILINGS of the plane.

dipyramid A polyhedron formed by joining two congruent pyramids base
to base.

direct Describing a TRANSFORMATION that does not change ORIENTATION.

directed angle An angle with a preferred direction of rotation, either
clockwise or counterclockwise, from one ray, called the initial side,
to the other ray, called the terminal side. The measure of a directed
angle can be positive or negative.

directed edge An edge of a GRAPH that has a given orientation or preferred
direction from one endpoint to the other.

directed graph *See* DIGRAPH.

directed magnitude A measurement with a positive or negative sign associated with it depending on the orientation of the object measured. For example, the directed magnitude of the length of segment *AB* is the negative of that of segment *BA* and the directed magnitude of the measure of a clockwise angle is negative.

directed segment A segment with a preferred direction from one endpoint, its initial point, to the other endpoint, its terminal point. The directed segment *AB* has initial point *A* and terminal point *B*.

direction cosine The cosine of an angle between a vector and a coordinate axis. A vector in an *n*-dimensional coordinate space has *n* direction cosines.

direction field An assignment of a direction to each point in space. A direction field is a VECTOR FIELD with the property that each vector has unit length.

direction vector A VECTOR of unit length that is used to give a direction.

directional derivative A measure of how much a SCALAR FUNCTION changes along a given direction in the domain space.

director circle For an ellipse or parabola, the circle that is the locus of the intersections of all pairs of tangents of the ellipse or parabola that are perpendicular to each other.

director cone For a RULED SURFACE and a fixed point, the cone of all lines through the point that are parallel to a GENERATOR of the ruled surface. For example, the director cone of a right circular cylinder is a line and the director cone of a cone is a cone.

directrix *See* CONIC SECTION.

Dirichlet domain *See* VORONOI REGION.

Dirichlet region *See* DIRICHLET TILING.

Dirichlet tessellation *See* DIRICHLET TILING.

Dirichlet tiling For a nonempty collection of mutually disjoint sets in the plane, the DIRICHLET tiling is the subdivision of the plane into regions, called Dirichlet regions, each containing one of the sets, such that each point in the region is closer to that set than to any other of the sets. If each set contains a single point, the Dirichlet tiling is a VORONOI DIAGRAM.

disconnected Describing a shape, space, or graph that is not CONNECTED.

disconnecting edge An edge of a GRAPH that, when removed, breaks the graph into two COMPONENTS.

discontinuity A point in the domain of a function where the function is not CONTINUOUS. A break or gap appears in the graph of the function at a discontinuity. For example, $y = 1/x$ has a discontinuity at $x = 0$.

discrete An ISOMETRY with the property that a point and its image are farther apart than some given positive number. For example, a translation is discrete, but a rotation or reflection can never be discrete.

discrete dynamical system A DYNAMICAL SYSTEM governed by a function that gives the transition from one state to another. Long-term behavior of the dynamical system is studied by looking at iterations of the function.

discrete geometry The study of finite sets of points, lines, circles, planes, or other simple geometric objects.

discrete topology The TOPOLOGY on a given set in which every subset of the given set is an OPEN SET.

disjoint Describing two or more sets that do not have any elements in common.

disjunction A statement formed by joining two statements with the word *or*.

disphenoid A polyhedron with four congruent triangular faces and with congruent solid angles.

displacement An ISOMETRY that preserves ORIENTATION.

dissection The partitioning of a region into regions that are disjoint or intersect only along their boundaries. *See also m-*DISSECTION.

dissection tiling A MONOHEDRAL TILING with PROTOTILE that is a rearrangement of the pieces of a given polygon.

distance The measure of separation between two geometric objects: two points, a point and a line, and so on. Distance is given as a non-negative real number. The distance between two figures is the length of the shortest segment connecting a point of one of the figures to a point of the other. Thus, the distance between a point and a line is the length of the perpendicular segment from the point to the line; the distance between two parallel lines or two skew lines is the length of a perpendicular segment from one to the other; and the distance between a point and a plane is the length of a perpendicular segment from the point to the line.

distinct Different from; not equal to. Two objects are not assumed to be distinct unless specifically stated to be so.

divergence The measure of how much fluid is entering or leaving a system at a point of a VECTOR FIELD representing fluid flow.

divide externally The point *P* divides segment *AB* externally if *P* is on the line through *A* and *B* but is not between *A* and *B*. The ratio in which *P* divides *AB* externally is *AP/PB*.

divide internally The point *P* divides segment *AB* internally if *P* is on segment *AB* between *A* and *B*. The ratio in which *P* divides *AB* internally is *AP/PB*.

divided proportionately Two segments are divided proportionately if they are cut so that the resulting segments of one are proportional to the resulting segments of the other. The segments *AB* and *CD* with points *X* on *AB* and *Y* on *CE* are divided proportionately if *AX/XB = CY/YD*.

divider dimension The FRACTAL DIMENSION *d* of a curve determined by the equation $L = c \, l^{1-d}$ where *L* is the divider length of the curve, *c* is a constant, and *l* is the step length. It is not necessary to know *c* to determine *d* if measurements are made using two or more different step lengths.

divider measure The length of a curve using DIVIDER MEASUREMENT.

divider measurement Measurement of a curve using dividers set at a fixed distance, called the step-length. The divider is used to approximate a short arc of the curve by a segment, called a divider step, whose length is equal to the step-length. The length of the curve is the step-length times the number of divider steps used to traverse the curve.

divider step *See* DIVIDER DIMENSION.

dividers *See* COLLAPSING COMPASS.

dividing point A point in the interior of a segment. It divides the segment into two segments.

divine proportion *See* GOLDEN RATIO. This name was given by Fra Luca Pacioli.

divine section *See* GOLDEN SECTION.

D_n The DIHEDRAL GROUP of ORDER *n*.

dodecadeltahedron A polyhedron with 12 faces and eight vertices.

dodecagon A polygon with 12 sides.

dodecagram The star polygon created from the dodecagon. It has 12 points.

dodecahedron A polyhedron with 12 congruent faces, each a regular pentagon. Sometimes, a dodecahedron is any polyhedron with 12 faces.

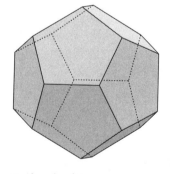

Dodecahedron

dog saddle A surface with a SADDLE POINT having four downward sloping regions (for the dog's legs) separated by four upward sloping regions.

domain (1) The set of points which are the inputs of a function. (2) A CONNECTED set.

dot pattern A pattern whose MOTIF is a single point.

dot product For two VECTORS, the product of their magnitudes times the cosine of the angle between them (which must be between 0° and 180°). The dot product of vectors **v** and **w** is denoted **v**·**w**.

double cone The surface swept out by a line rotated about an axis that intersects the line. The vertex of the cone is the intersection of the line with the axis.

double cusp *See* POINT OF OSCULATION.

double elliptic geometry An ELLIPTIC GEOMETRY in which any two lines intersect at two points. The surface of a sphere is a double elliptic geometry since any two great circles intersect at two distinct points.

double point (1) A point where a curve intersects itself or where there is a CUSP. (2) *See* FIXED POINT OF A FUNCTION.

double ratio *See* CROSS RATIO.

double-rectangular tetrahedron *See* QUADRIRECTANGULAR TETRAHEDRON.

double root A root a of a polynomial equation that has exactly two factors of the form $(x - a)$.

double surface A one-sided NONORIENTABLE surface such as the MÖBIUS BAND or KLEIN BOTTLE.

double torus A sphere with two HANDLES attached.

doubly asymptotic triangle In HYPERBOLIC GEOMETRY, a line together with a LEFT-SENSED PARALLEL and a RIGHT-SENSED PARALLEL. It has only one vertex.

doubly connected 2-connected; *see* n-CONNECTED.

doubly ruled surface A RULED SURFACE that can be generated by a moving line in two different ways. The only doubly ruled surfaces are the plane, the HYPERBOLOID of one sheet, and the HYPERBOLIC PARABOLOID.

dragon curve A fractal curve created from a right triangle through the iteration of a series of DILATIONS, rotations, and translations.

drop a perpendicular To construct a perpendicular from a given point to a given straight object.

Droz-Farny circles There are two congruent Droz-Farny circles for any pair of ISOGONAL CONJUGATE POINTS, P and Q, belonging to a given triangle. To construct one of the circles, drop an altitude from point P to a side of the triangle; with the foot of this altitude as center, construct a circle passing through Q; this circle intersects the side of the triangle in two places; the circle centered at P passing through these two points is one Droz-Farny circle. The other is constructed similarly starting with point Q.

dual *See* DUALITY.

dual polyhedron A POLYHEDRON whose faces are the vertex figures of another polyhedron.

dual statement A statement obtained by switching the words *point* and *line* in two-dimensional PROJECTIVE GEOMETRY, or obtained by switching the words *point* and *plane* in three-dimensional projective geometry. These dual statements are called the plane dual and the space dual, respectively. In projective geometry, the truth of a statement implies the truth of its dual statement.

dual tiling A tiling whose edges are the VERTEX FIGURES of a given tiling. It is formed by placing a vertex in every face of the given tiling and connecting two of these vertices by an edge if they are in adjacent faces of the given tiling.

duality In general, a pairing or relationship between objects of one type with objects of another type. Two objects related by a duality are said to be dual to one another.

duality, principle of The principle that if the terms *point* and *line* (or *point* and *plane* in a three-dimensional projective space) are interchanged in an axiom or theorem in PROJECTIVE GEOMETRY, the resulting statement is still true.

duodecagon *See* DODECAGON.

duodecahedron *See* DODECAHEDRON.

duplication of the cube The problem of finding a cube with a volume twice that of a given cube. One of the three famous problems of antiquity, this problem is impossible to solve with compass and straightedge.

dust A fractal consisting of a discrete set of points; also called FATOU dust, CANTOR dust, or fractal dust.

dyad An ordered pair of real numbers.

dyadic set *See* CANTOR SET.

Dymaxion The map formed by projecting the surface of the sphere onto an ICOSAHEDRON or a CUBOCTAHEDRON. It was first used by BUCKMINSTER FULLER.

dynamic geometry Trademark term for GEOMETER'S SKETCHPAD interactive geometry software.

dynamic symmetry A method of composing a painting based on the GOLDEN SECTION that was developed by Jay Hambidge in the early 20th century.

dynamical system A mathematical description, usually by a DIFFERENCE EQUATION or DIFFERENTIAL EQUATION, of a physical system that changes with respect to time.

e The irrational number approximately equal to 2.71828459. It is used as the base for natural logarithms.

eccentric circle A larger circle whose center moves along a smaller circle. The ancient Greeks used eccentric circles to describe the motion of the planets.

eccentricity *See* CONIC SECTION.

edge A line or segment that has a special role depending on the context. The line determining a half-plane is its edge. In a polyhedron, an edge is a segment connecting two vertices and incident to two faces. In a graph, an edge is a line connecting two nodes.

edge sphere A sphere tangent to all the edges of a REGULAR POLYHEDRON.

edge-to-edge tiling A tiling by polygons with the property that the intersection of any two tiles is completely contained in one edge of each of the tiles.

Egyptian triangle A right triangle with sides equal to 3, 4, and 5.

eigenbasis A BASIS of the domain of a LINEAR TRANSFORMATION that consists of EIGENVECTORS of the transformation.

eigenspace For a given square matrix, the vector space of all EIGENVECTORS having the same EIGENVALUE.

eigenvalue The SCALAR that an EIGENVECTOR is multiplied by when the eigenvector is multiplied by a matrix.

eigenvector A VECTOR whose direction is left unchanged or reversed when multiplied by a given square matrix.

elation A PERSPECTIVE COLLINEATION with the property that its center lies on its axis.

elbow A joint in a ROBOT ARM.

element An object contained in a set. The notation $x \in S$ means that x is an element of the set S.

elementary curve A curve in three-dimensional space with the property that every point is contained in a CLOSED BALL whose intersection with the curve is a DEFORMATION of a CLOSED SEGMENT.

elementary geometry *See* EUCLIDEAN GEOMETRY.

elementary surface A surface in three-dimensional space with the property that every point is contained in a CLOSED BALL whose intersection with the surface is a DEFORMATION of a CLOSED DISC.

elevation The height of a point above a given surface.

ellipse The intersection of a cone with a plane that meets the cone in a CLOSED CURVE. Equivalently, an ellipse is the locus of all points whose distances from two points, called the foci of the ellipse, have a constant sum.

ellipsoid of revolution The surface generated when an ELLIPSE is rotated about one of its axes. Its cross sections are circles or ellipses. An ellipsoid has three AXES OF ROTATION: the major axis (the longest); the mean axis; and the minor axis (the shortest).

elliptic axiom The AXIOM that states that two lines always intersect. This is used instead of Euclid's fifth postulate to define ELLIPTIC GEOMETRY.

elliptic curve A curve given by the equation $y^2 = ax^3 + bx^2 + cx + d$. If a, b, c, and d are rational, the curve is a rational elliptic curve.

elliptic cylinder A cylinder whose base is an ellipse.

elliptic geometry A non-Euclidean geometry in which Euclid's parallel postulate is replaced by the ELLIPTIC AXIOM which asserts that any two lines intersect. A model for elliptic geometry is the sphere.

elliptic integral An INTEGRAL that can be used to find the ARC LENGTH of an ELLIPSE or the position of a pendulum.

elliptic paraboloid A QUADRIC SURFACE that has elliptical cross sections. It is given by the equation $x^2/a^2 + y^2/b^2 = z$.

elliptic point A point on a space curve where, locally, the curve lies on one side of the tangent line; or a point on a surface where, locally, the surface lies on one side of the tangent plane. At an elliptic point, the CURVATURE is nonzero.

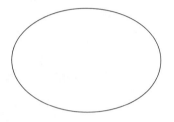

Ellipse

elliptic projectivity *See* PROJECTIVITY.

elliptic rotation An EQUIAFFINITY of the form $f(x, y) = (x \cos \theta - y \sin \theta,$ $x \sin \theta + y \sin \theta)$ for any θ.

elliptic umbilic An ELLIPTIC POINT that is an UMBILIC.

elongated dodecahedron A space-filling ZONOHEDRON with 12 faces: four equilateral hexagons and eight rhombs.

elongated rhombic dodecahedron A space-filling ZONOHEDRON with 12 faces, four hexagons, and eight rhombs.

elongated square gyrobicupola *See* PSEUDORHOMBICUBOCTAHEDRON.

elongation A transformation of the plane given by the equations $f(x, y) = (x, ky)$ or $f(x, y) = (kx, y)$ where $k > 1$.

embed To map a set into a space by a one-to-one function.

empire The set of all FORCED tiles for a given patch of tiles in a given tiling.

empty set The set containing no elements, denoted $\varnothing$.

enantiomers Two CHIRAL molecules that are mirror images of one another.

enantiomorph A figure having a MIRROR IMAGE that is not congruent to itself.

end effector A ROBOT HAND that can grip, weld, drill, paint, or perform some other specific task.

end-centered lattice A LATTICE consisting of the points of a given lattice together with the center points of two opposite faces of each LATTICE UNIT.

endecagon *See* HENDECAGON.

endpoint A point on a segment or ray that is not between any two other points on the segment or ray.

enneacontagon A polygon with 90 sides.

enneagon *See* NONAGON.

enneahedron A polyhedron with nine faces.

enumerative geometry A study of the number of parameters needed to define various families of geometric objects.

enunciation For a theorem, the statement of what is known and what is to be proved.

envelope The set of all lines tangent to a given curve or the set of all planes tangent to a given surface.

envelope of normals *See* EVOLUTE.

epicycle (1) In the construction of an EPICYCLOID, the circle that rolls along the fixed circle. (2) A circle whose center travels along the circumference of another circle. The ancient Greeks used epicycles to model the motion of the planets with respect to the Earth.

epicycloid The locus of a point fixed on the circumference of a circle that rolls without sliding on the outside of a circle.

epigraph For a real-valued function, the set $\{(x, y) \mid y \geq f(x)\}$. It consists of the graph of f together with all points above the graph.

epitrochoid The locus of a point fixed on a ray from the center of a circle that rolls without slipping on the outside of another circle.

epitrochoidal The locus of a point fixed on a ray from the center of a circle that rolls without slipping on the outside of another circle such that the planes of the two circles always meet at the same angle.

equal The same as.

equal sign The symbol =, read "is equal to" or "is the same as."

equal tilings Two tilings that can be related by a SIMILARITY TRANSFORMATION.

equality The property of being the same as.

equant A point close to the center of a circle, used as the center of an ECCENTRIC circle in PTOLEMY'S EPICYCLE model of planetary motion.

equation A mathematical expression stating that two quantities are equal. For example, $2 + 3 = 5$ is an equation.

equator *See* GREAT CIRCLE.

equatorial circle For a given point on a sphere, the GREAT CIRCLE that passes through that point.

equatorial polygon A polygon that is the intersection of a polyhedron with a plane through the center of the polyhedron with the property that its vertices are vertices of the polyhedron and its sides are edges of the polyhedron. Not every polyhedron has an equatorial polygon.

equiaffine collineation *See* EQUIAFFINITY.

equiaffinity An AFFINE TRANSFORMATION that preserves area or volume.

equiangular hyperbola *See* RECTANGULAR HYPERBOLA.

equiangular polygon A polygon with all angles congruent to one another.

equiangular spiral *See* LOGARITHMIC SPIRAL.

equiareal Area-preserving. On an equiareal map, regions have the same relative area that they do on Earth.

equichordal point A point in a region with the property that all segments through the point that join two boundary points of the region have equal length.

equidecomposable Describing two regions with the property that each can be cut into pieces that can be rearranged to give a region congruent to the other. Any two polygons with equal area are equidecomposable.

equidistant Describing an object having equal distances to each of a collection of objects.

equidistant curve An equidistant curve is a generalization of circle in HYPERBOLIC GEOMETRY. It is the set of all points whose perpendicular distance to a given axis is constant.

equidistant projection A PROJECTION that preserves the distance between one or two fixed points and every other point.

equidistant surface An equidistant surface is a generalization of sphere in HYPERBOLIC GEOMETRY. It is the set of all points whose perpendicular distance to a given plane is constant.

equiform geometry *See* SIMILARITY GEOMETRY.

equiform transformation *See* SIMILARITY.

equilateral hyperbola *See* RECTANGULAR HYPERBOLA.

equilateral polygon A polygon with all sides congruent to one another.

equilateral triangle A triangle with all sides congruent to one another.

equilibrium point A fixed point of a DYNAMICAL SYSTEM.

equitransitive Describing a tiling by regular polygons with the property that any two congruent tiles are equivalent to each other.

equivalence class The set of all elements EQUIVALENT to a given element with respect to a given EQUIVALENCE RELATION.

equivalence relation A RELATION that is REFLEXIVE, SYMMETRIC, and TRANSITIVE. Equality, congruence, and similarity are equivalence relations.

equivalent Equal or related by an EQUIVALENCE RELATION. For a tiling, two vertices, VERTEX FIGURES, or tiles are equivalent if one can be mapped to the other by a symmetry of the tiling.

Erlanger Program The organization of geometries according to their GROUPS of TRANSFORMATIONS that was proposed by FELIX KLEIN. He stated that each geometry is the study of properties left invariant by a group of transformations of the space of the geometry.

escape set All points that are mapped further and further away from their original position by the iterations of a DYNAMICAL SYSTEM.

escribed circle *See* EXCIRCLE.

estimation The process of finding a number that is close to a measurement or computed value.

Euclidean axioms The axioms given by EUCLID in the *Elements.*

Euclidean geometry The geometry described by the axioms of EUCLID in the *Elements.*

Euclidean group The GROUP of all ISOMETRIES of a EUCLIDEAN SPACE.

Euclidean norm The length of a VECTOR. The norm of **v** is denoted $\|\mathbf{v}\|$.

Euclidean space A space satisfying the EUCLIDEAN AXIOMS.

Euclidean tools Compass and straightedge.

Euler brick A CUBOID whose length, width, and height are different and whose edges and face diagonals have integer lengths.

Euler characteristic The EULER NUMBER or a generalization of the Euler number.

Euler circuit A LOOP of EDGES in a GRAPH that includes each edge once and only once.

Euler formula (1) The formula $e^{i\theta} = \cos\theta + i\sin\theta$. (2) For a simple connected polyhedron, $V - E + F = 2$, where V is the number of vertices of the polyhedron, E is the number of edges, and F is the number of faces.

Euler formulas The formulas that determine the FOURIER COEFFICIENTS belonging to a given periodic function.

Euler line The line passing through the CIRCUMCENTER, the ORTHOCENTER, and the CENTROID of a triangle.

Euler number For a polyhedron, the number of faces minus the number of edges plus the number of vertices. The Euler number of a simply connected polyhedron is 2.

Euler points of a triangle The midpoints of the segments connecting the ORTHOCENTER of a triangle to its vertices.

Euler's angles The angles between the axes of two different three-dimensional coordinate systems belonging to the same underlying space.

Euler's spiral *See* SPIRAL OF CORNU.

Euler's triangle *See* SPHERICAL TRIANGLE.

Euler-Poincaré characteristic For a surface *M*, the Euler-Poincaré characteristic, denoted c(*M*), is the number of faces minus the number of edges plus the number of vertices for any map on the surface. For a surface embedded in three-dimensional space, the Euler-Poincaré characteristic is the number of local maxima plus the number of local minima minus the number of saddle points.

Euler-Poincaré formula A formula that gives the Euler number of a finite simplicial complex as the alternating sum of the BETTI NUMBERS of the complex. In the alternating sum, each Betti number is multiplied by 1 if it is even-dimensional and by –1 if it is odd-dimensional.

Eulerian polyhedron A SIMPLY CONNECTED polyhedron; so called because it satisfies the EULER FORMULA.

eutactic star The ORTHOGONAL PROJECTION of a CROSS in higher dimensional space onto three-dimensional space.

even face A face of a polyhedron that has an even number of sides.

even function A function from the real numbers to the real numbers whose graph has mirror symmetry across the *y*-axis. Equivalently, a function $y = f(x)$ with the property that $f(-x) = f(x)$. Examples of even functions are $y = x^4$ and $y = \cos x$.

even isometry *See* DIRECT ISOMETRY.

even order horizon *See* SHADOW SEGMENT.

even permutation A permutation that is the product of an even number of TRANSPOSITIONS.

even vertex A vertex of a GRAPH incident to an even number of edges.

evolute The evolute of a curve consists of the CENTERS OF CURVATURE of all the points on the curve. The evolute of a surface is the surface consisting of all the centers of curvature in the PRINCIPLE DIRECTION at each point on the surface.

exactitude of a construction The total number of intersections that occur in a particular construction by compass and straightedge.

excenter The center of an EXCIRCLE. It is the intersection of the bisectors of two EXTERIOR ANGLES of a triangle and the bisector of the interior angle at the third vertex.

exceptional object In robotics, an object that does not allow a CLOSURE GRASP.

exceptional plane The plane of EXCEPTIONAL POINTS of a PERSPECTIVE PROJECTION. It is the plane through the center of projection parallel to the image plane.

exceptional point A point that does not have an image under a PERSPECTIVE PROJECTION. An exceptional point is on the plane through the center of projection parallel to the image plane.

excess In SPHERICAL or ELLIPTIC GEOMETRY, the ANGLE SUM of a triangle minus 180°.

excircle A circle outside a triangle that is tangent to one side of the triangle and to the two other sides once they have been extended. Every triangle has three excircles.

exhaustion, method of A method of computing the area or volume of a figure by approximating it with a sequence of inscribed polygons or polyhedra. This method was developed by the Greek mathematician EUDOXUS and later used by ARCHIMEDES.

existence axiom An AXIOM that states that a certain object or objects exist.

existence theorem A theorem whose conclusion is that a certain mathematical object exists.

exotic sphere A seven-dimensional manifold that is HOMEOMORPHIC to the seven-dimensional sphere but is not DIFFEOMORPHIC to it.

expansion A similarity for which the RATIO OF SIMILITUDE is greater than 1.

exponential decay A process modeled by the equation $y = ae^{-bx}$, with a and b constant, $b > 0$.

exponential growth A process modeled by the equation $y = ae^{bx}$, with a and b constant, $b > 0$.

exradius The radius of an EXCIRCLE.

exsecant For angle A, the exsecant is sec $A - 1$.

extended Euclidean plane The EUCLIDEAN PLANE with a LINE AT INFINITY added. Two lines parallel in the Euclidean plane meet at a point on the line at infinity in the extended Euclidean point.

extended proportion *See* CONTINUED PROPORTION.

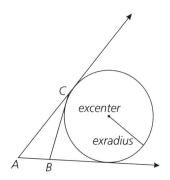

An excircle of triangle ABC

extension The extension of a segment or ray is the line containing it.

exterior The region outside a polygon, closed curve, or surface. It is the complement of the union of the figure and its interior.

exterior angle An angle formed by one side of a polygon and the extension of one of its adjacent sides.

exterior angle sum The sum of the measures of one EXTERIOR ANGLE at each vertex of a polygon. This sum is 360° for any convex polygon.

exterior cevian A segment from a vertex of a triangle to the extension of the opposite side. It does not pass through the interior of the triangle.

external Outside.

external bisector The bisector of an EXTERIOR ANGLE of a polygon.

external division of a line The division of a segment by a point collinear to the segment but not contained in it. For example if B is between A and P, then P divides the segment AB externally with ratio AP/PB.

external symmetry In physics, a global symmetry of SPACETIME GEOMETRY.

extrapolation The use of a straight line or other simple curve to approximate a point near, but not between, two given points on a more complex curve or surface.

extreme and mean ratio *See* GOLDEN RATIO.

extreme point A point in a CONVEX set that is not contained as an interior point of a segment in the set. When an extreme point is removed from a convex set, it is still convex.

extremes of a proportion The numbers a and d in the proportion $a/b = c/d$.

extrinsic curvature The curvature of a curve or surface with respect to measurements taken in the space containing the curve or surface.

extrinsic distance The distance in the space containing a surface between two points on the surface. *See also* INTRINSIC DISTANCE.

face (1) One of the polygons making up a polyhedron. Its vertices are vertices of the polyhedron and its sides are edges of the polyhedron. (2) A region bounded by edges of a PLANAR GRAPH that contains no edges of the graph.

face angle An angle of a face of a polyhedron.

face coloring A coloring of each region of a MAP or GRAPH.

face diagonal A segment that joins two vertices of a polyhedron and lies on a face of the polyhedron.

face of a simplex A simplex of lower dimension that is contained in a simplex.

face-centered lattice A LATTICE containing the points of a given lattice together with the center points of the faces of each LATTICE UNIT.

face-regular compound A COMPOUND POLYHEDRON whose faces lie on the faces of a regular polyhedron. For example, the stella octangula is a face-regular compound.

facet (1) The higher dimensional analogue of face; it is an $(n-1)$-dimensional polytope that is part of the boundary of an n-dimensional polytope. (2) To remove a pyramidal-shaped piece from a polyhedron.

factor knot One of the KNOTS making up a COMPOSITE KNOT.

factorize To find all the factors of a product; thus 15 can be factorized as 3×5.

Fagnano's problem The problem of finding the inscribed triangle with smallest possible perimeter in an acute-angled triangle. The solution is the ORTHIC TRIANGLE.

fallacy Invalid or incorrect reasoning.

family of curves A collection of curves, usually having some common feature in their definition. For example, all circles with the same center form a concentric family of circles.

Fano configuration A CONFIGURATION consisting of seven lines and seven points in which each line contains three points.

Fano plane A PROJECTIVE PLANE in which the three diagonal points of every quadrangle are collinear.

Fano's axiom The AXIOM that says that the three diagonal points of a complete quadrangle are never collinear.

Fano's geometry A FINITE GEOMETRY consisting of seven lines and seven points in which each line contains three points.

feature space A vector space used in computer vision in which each vector represents an object or potential object. The components of a vector correspond to the measurements, or features, of the object represented by the vector.

Fedorov solid *See* PARALLELOHEDRON.

Feigenbaum's number The number governing period doubling of DYNAMICAL SYSTEMS. It is the ratio of consecutive increases in the value of a parameter that will produce BIFURCATION of the dynamical system and is approximately 4.6692.

Fermat point The point inside a triangle with the property that segments connecting it to any two vertices of the triangle meet at an angle of 120°. *See* FERMAT'S PROBLEM.

Fermat prime A prime number of the form $2^{2^k} + 1$. The only Fermat primes known today are 3, 5, 17, 257, and 65,537.

Fermat's problem The problem of finding a point whose distances to the three vertices of a given acute-angled triangle have the smallest possible sum. The solution is called the FERMAT POINT of the triangle.

Feuerbach circle *See* NINE-POINT CIRCLE.

Fibonacci sequence The sequence of numbers 1, 1, 2, 3, 5, 8, 13, 21, . . . in which each number after the first two is the sum of the previous two numbers in the sequence.

fiber The PREIMAGE of an element under a mapping. *See also* FIBER BUNDLE.

fiber bundle A "twisted" Cartesian product. A fiber bundle consists of a total space, a base space embedded in the total space, and a fiber. Near each point in the base space, the total space looks like the Cartesian product of a region in the base space with the fiber, but globally the space may not be a Cartesian product. Thus, the MÖBIUS BAND is a fiber bundle with base space a circle and fiber a segment.

fifth postulate *See* PARALLEL POSTULATE.

figurate number An integer n with the property that n dots can be arranged to form a triangle, square, rectangle, or pentagon.

figure-eight knot A KNOT that appears similar to a figure eight.

filled Julia set *See* PRISONER SET.

fillet The region inside a square and outside a circle with center on a vertex of the square and with radius equal to a side of the square.

fine planning The aspect of planning a robot's path that takes into account unpredictable obstacles, such as other robots, met along the path.

finger-gaiting *See* REGRASPING.

finite geometry A geometry in which there are only finitely many points.

finite intersection property The property of an infinite collection of sets that any finite collection of the sets has nonempty intersection.

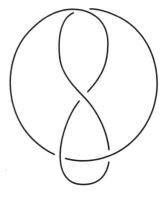

Figure-eight knot

finitely generated Having a finite basis.

first Brocard triangle The triangle inscribed in the BROCARD CIRCLE of a given triangle such that each vertex is the intersection of the Brocard circle with a ray from a vertex of the triangle to a BROCARD POINT of the triangle.

first curvature *See* CURVATURE.

first Lemoine circle For each side of a triangle, construct a parallel that passes through the SYMMEDIAN POINT. The first Lemoine circle passes through the six points of intersection of these parallels with the sides of the triangle.

fixed point A point whose image under a function or transformation is itself.

flag A set of objects that are mutually incident. For example, a vertex, a side that contains it, and a polygon that contains the side together form a flag.

flat *See* AFFINE SUBSPACE.

flat angle *See* STRAIGHT ANGLE.

flat pencil A PENCIL of lines that lies in a plane.

Flatland A plane on which imaginary two-dimensional creatures live, created by British clergyman E. A. ABBOTT in the satire *Flatland*.

flecnode A NODE of a curve at which one of the arcs of the curve has an inflection point.

flex A very small MOTION of a FRAMEWORK.

flex point *See* INFLECTION POINT.

flexion The DERIVATIVE of a derivative; acceleration. This is the term used by NEWTON.

flip *See* REFLECTION.

flow A DYNAMICAL SYSTEM, especially one given by a vector field in Euclidean space.

flux The rate of change with respect to time of the mass contained within a given region.

fluxion The term used by ISAAC NEWTON for DERIVATIVE.

focal point An EQUILIBRIUM POINT of a DYNAMICAL SYSTEM that is the limit of a spiral trajectory of the system.

focal radius A segment connecting a point on a conic section to a focus of the conic section.

focal surface The evolute of a surface. *See* EVOLUTE.

focus *See* CONIC SECTION.

fold A singularity of a function from a surface to a surface that occurs when the domain surface gets "folded" before it is mapped to the range. Each point on the fold has one inverse image; each point on one side of the fold has two inverse images; and each point on the other side of the fold has no inverse images. For example, a fold results at the equator when a sphere is projected onto a plane.

fold point A singularity where there is a fold.

foliation A collection of disjoint pieces, called the leaves of the foliation, all of the same dimension, whose union is a given manifold.

folium of Descartes The curve given by the equation $x^3 + y^3 = 3axy$, where a is a constant. It has a single loop and is asymptotic to the line $y = -x - a$. The area of the loop is equal to the area between the curve and its asymptote.

foot The point where a perpendicular line intersects the straight object to which it is perpendicular or where a CEVIAN of a triangle intersects a side of the triangle.

force field A vector field whose vectors represent a force.

forced tile In a tiling, a tile whose shape and orientation are completely determined by tiles already in place.

forest A GRAPH which does not contain any circuit; a collection of TREES.

forward kinematic problem Calculation of the position and orientation of a robot hand in terms of the lengths of the links on the robot arm and the angles between them. This is an important problem in robotics.

four-color problem The problem of determining whether or not any map on a given surface has a map coloring using at most four colors. For the plane, four colors suffice for all maps and the result is known as the four-color theorem.

Fourier coefficients The coefficients of the terms in a FOURIER SERIES.

Fourier series The TRIGONOMETRIC SERIES that most closely approximates a given periodic function.

fourth harmonic For a given set of three COLLINEAR points, a point that, together with the three given points, forms a HARMONIC RANGE.

fourth proportional For numbers a, b, and c, with $b \neq 0$, the number x such that $a/b = c/x$.

fractal A geometrical figure in which a motif or shape is repeated at ever-diminishing scales. An example of a fractal is the KOCH SNOWFLAKE.

fractal dimension Fractal dimension measures how dense a fractal is in the space containing it. There are several different ways to define fractal dimension, including SELF-SIMILAR FRACTAL DIMENSION, BOX DIMENSION, DIVIDER DIMENSION, and HAUSDORFF DIMENSION.

frame An ordered set of ORTHOGONAL UNIT VECTORS having a common tail. A frame can be used as the set of BASIS vectors for a vector space.

framework A system of rigid rods—represented by segments—that are attached by hinges—represented by vertices—and embedded in two- or three-dimensional space.

Franciscan cowl *See* BRIDE'S CHAIR.

Fraser spiral A tiling of the plane by polygons that has a spiral-like appearance.

free action A TRANSFORMATION with no fixed points.

free path A path for a robot on which there are no obstacles.

free vector A vector whose initial point can be anywhere.

frequency A measure of how complex a GEODESIC POLYHEDRON is. *See* ALTERNATE METHOD.

frieze group The GROUP of SYMMETRIES of a BAND ORNAMENT or frieze pattern; there are only seven possible such groups.

frieze pattern *See* BAND ORNAMENT.

frontier *See* BOUNDARY.

frustum That part of a geometric solid between two parallel planes, especially that part of a cone or pyramid left when the top is cut off by a plane parallel to the base.

Fuhrmann circle For a triangle, the circle whose diameter is the segment joining the ORTHOCENTER and the NAGEL POINT of the triangle.

Fuhrmann triangle Construct the midpoints of the arcs of the CIRCUMCIRCLE subtended by the sides of a given triangle. The vertices of the Fuhrmann triangle are the reflections of each of these midpoints across the nearest side.

full circle (full angle) An angle equal to 360°.

fullerene A polyhedron whose faces are pentagons and hexagons, with three faces meeting at each vertex, and with exactly 12 pentagonal faces. A fullerene molecule can be modeled by a fullerene polyhedron; it consists of carbon atoms corresponding to the vertices of a fullerene with chemical bonds corresponding to the edges of the fullerene.

function A rule that assigns to each point in one set (the domain of the function) a point in another set (the range of the function).

function space A TOPOLOGICAL SPACE in which each point is a function with the same domain and range.

fundamental domain A CONNECTED set whose images under a group of TRANSLATIONS of a space are disjoint and cover the space.

fundamental group The group of LOOPS in a given space all starting and ending at the same point. The group operation is the concatenation of loops. Two loops are equivalent if one can be DEFORMED into the other within the given space. The fundamental group provides a way to measure the holes in a topological space; the fundamental group of the sphere is trivial and the fundamental group of the torus has two GENERATORS.

fundamental lattice *See* STANDARD LATTICE.

fundamental parallelepiped A three-dimensional fundamental domain.

fundamental parallelogram A two-dimensional fundamental domain.

fundamental region *See* FUNDAMENTAL DOMAIN.

fundamental triangle The first quadrant of the PROJECTIVE PLANE. It is the set of all points in the projective plane with homogeneous coordinates (x, y, z) such that $x \geq 0$ and $y \geq 0$.

gamma point *See* ULTRA-IDEAL POINT.

gauge field An assignment of a GAUGE GROUP to each point in a MANIFOLD.

gauge group A SYMMETRY GROUP that acts locally on a small region of a MANIFOLD.

gauge theory A theory of physics that uses GAUGE FIELDS to describe forces.

Gauss map A map from the points of an oriented plane curve to the unit circle; the image of a point is the tip of a unit tangent displaced so its tail is at the origin. Also, a map from the points of an oriented space curve or an oriented surface to the unit sphere; the image of a point is the tip of a unit normal vector displaced so its tail is at the origin.

Gaussian coordinates *See* SURFACE COORDINATES.

Gaussian curvature A measure of the intrinsic curving of a surface at a point, given as the product of the two PRINCIPLE CURVATURES. It is positive at elliptic points, negative at saddle points, and 0 at parabolic points.

Gaussian image The image of a curve or surface under the GAUSS MAP.

general cone of the second order The surface consisting of segments connecting every point of a nondegenerate conic to a fixed noncoplanar point.

general position Not in any fixed pattern; points are in general position if no three are collinear and lines are in general position if no two are coplanar.

generating line *See* GENERATOR OF A SURFACE.

generating parallelogram *See* PERIOD PARALLELOGRAM.

generating region *See* FUNDAMENTAL DOMAIN.

generator of a surface A line lying entirely on a surface. A system of generators is a collection of nonintersecting generators that completely cover a surface.

generators of a group A set of elements that will give all elements of the group through repeated application of the group operation. For example, generators of the group of symmetries of a square are a 90° rotation and a reflection.

generatrix *See* GENERATOR OF A SURFACE.

genus A measure of how connected a surface is. It is the number of different nonintersecting SIMPLE CLOSED CURVES that can be removed from a surface without disconnecting it. The SPHERE has genus 0 and the TORUS has genus 1.

genus of a knot The smallest possible genus of a SEIFERT SURFACE for the KNOT. It is a KNOT INVARIANT.

geodesic The shortest path between any two points on a surface. A geodesic is intrinsically straight.

geodesic curvature *See* INTRINSIC CURVATURE.

geodesic distance *See* INTRINSIC DISTANCE.

geodesic mapping A function that maps GEODESICS onto geodesics.

geodesic polyhedron A polyhedron that can be inscribed in a sphere.

geodesy The accurate measurement of the Earth's size and shape and the determination of the position of specified points on the Earth's surface.

geodetic curve A GEODESIC on the surface of the Earth.

geodetic surveying Surveying that takes into account the curvature of the earth's surface.

geographical information systems Computer procedures for storing, analyzing, and displaying geographical and geophysical data.

geoid The shape of the Earth.

Geometer's Sketchpad Computer software that allows the user to create, manipulate, and measure geometric figures.

geometria situs A name for TOPOLOGY used in the 19th century.

geometric mean For two numbers a and b, the geometric mean is $\sqrt{ab}$, which is also called the mean proportional of the two numbers. For n numbers $a_1, a_2, \ldots, a_n$, the geometric mean is $\sqrt{a_1 a_2 \ldots a_n}$.

geometric multiplicity The geometric multiplicity of an EIGENVALUE is the DIMENSION of its EIGENSPACE.

geometric probability The application of geometry to compute probabilities. If a sample space can be represented by a region in a plane with an event represented by a subregion, the probability of the event is the ratio of the area of the subregion to the area of the whole region.

geometric sequence An infinite sequence of the form $a, ar, ar^2, \ldots$.

geometric series An infinite sum of the form $a + ar + ar^2 + \ldots$.

geometrical algebra The use of geometry, especially the application of areas, to solve algebraic equations.

geometrography A procedure for comparing the efficiency of different methods of constructing a geometric figure with compass and straightedge by counting the number of operations used by each method.

geometry (1) The mathematical study of shapes, forms, their transformations, and the spaces that contain them. (2) A specific AXIOMATIC SYSTEM that studies shape, form, transformations, and spaces, such as EUCLIDEAN GEOMETRY or PROJECTIVE GEOMETRY, is referred to as a geometry.

geometry of numbers The use of geometry and geometrical reasoning in NUMBER THEORY.

Gergonne point The point of concurrency of the lines connecting each vertex of a triangle to the point where the opposite side meets the incircle.

Gergonne triangle The triangle whose vertices are the points of contact of the incircle of a given triangle.

gift wrapping algorithm An algorithm for finding the CONVEX HULL of a given set of points.

girth A measure of the size of a shape. The girth of a solid is the perimeter of any one of its orthogonal parallel projections onto a plane. The girth of a GRAPH is the number of edges in the shortest circuit contained in the graph.

GIS *See* GEOGRAPHICAL INFORMATION SYSTEMS.

given That which is assumed in the context of a particular statement, problem, or theorem.

glide reflection An ISOMETRY that is the composite of a REFLECTION with a TRANSLATION along the mirror line of the reflection.

glide symmetry Symmetry with respect to a GLIDE REFLECTION.

glissette The locus of a point on a curve that slides on or between two fixed curves.

global Pertaining to or valid for the whole of a curve, surface, shape, or space.

global positioning system A system that uses vectors and geometry to determine an individual's position on the Earth from data transmitted by satellites.

gnomon Anything shaped like a carpenter's square. For example, the vertical part of a sundial or a region added to a polygon that results in a similar polygon are gnomons.

gnomonic projection *See* CENTRAL PROJECTION.

golden dodecahedron A DODECAHEDRON that has removed from each face the small pentagon bounded by the diagonals on each face.

golden mean *See* GOLDEN SECTION.

golden number *See* GOLDEN SECTION.

golden parallelogram A parallelogram with sides in the golden ratio and acute angles equal to 60°.

golden ratio *See* GOLDEN SECTION.

golden rectangle A rectangle for which the ratio of length to width is the GOLDEN RATIO.

golden section The cutting of a segment *AB* by a point *P* such that *AB/AP = AP/BP*. This ratio is called the golden ratio; it is equal to $(1 + \sqrt{5})/2$ and is denoted τ or Φ.

golden spiral The EQUIANGULAR SPIRAL $r = \tau^{2\theta}/\pi$, where τ is the GOLDEN RATIO.

gon A unit of angle measure; 400 gons are equal to 360°. This unit is used in surveying and aircraft navigation.

GPS *See* GLOBAL POSITIONING SYSTEM.

grad *See* GON.

grade *See* GON.

gradient A VECTOR pointing in the direction of maximum increase of a SCALAR FIELD.

gradient mapping The gradient mapping is a function associated with a smooth function that takes a point in the domain of the function to the GRADIENT vector of the function at that point.

graph (1) For a function $y = f(x)$, the graph is the set of all points in the Cartesian plane of the form $(x, f(x))$. For a function $z = f(x, y)$, the graph is the set of all points in three-dimensional space of the form $(x, y, f(x, y))$. (2) A set of points, called vertices or nodes, together with a set of segments or arcs, called edges or branches, whose endpoints belong to the set of vertices.

graph theory The study of the properties of GRAPHS formed of vertices and edges.

grasp A placement of the fingers of a robot hand on an object, especially a placement that holds the object in equilibrium.

great circle The intersection of a sphere with a plane passing through its center.

great dodecahedron A NONCONVEX polyhedron formed by 12 regular pentagons that pass through one another, five meeting at each vertex. It is a KEPLER-POINSOT SOLID.

great icosahedron A NONCONVEX polyhedron with 44 faces: 20 triangles, 12 pentagrams, and 12 decagrams. Its DUAL is the great stellated dodecahedron, and it is a KEPLER-POINSOT SOLID.

great rhombic triacontahedron A NONCONVEX ZONOHEDRON which is the DUAL of the GREAT ICOSIDODECAHEDRON. It is also called the great stellated triacontahedron.

great rhombicosidodecahedron A SEMIREGULAR POLYHEDRON with 62 faces: 30 squares, 20 hexagons, and 12 decagons.

great rhombicuboctahedron A SEMIREGULAR POLYHEDRON with 12 square faces, eight hexagonal faces, and six octagonal faces.

great stellated dodecahedron A NONCONVEX polyhedron formed by 12 regular pentagrams that pass through one another and meet three at each vertex. It has 60 triangular faces and can be formed by STELLATING an ICOSAHEDRON with triangular pyramids on each face. It is one of the KEPLER-POINSOT SOLIDS.

great stellated triacontahedron *See* GREAT RHOMBIC TRIACONTAHEDRON.

Grebe's point *See* SYMMEDIAN POINT.

Greek cross A cross-shaped region made up of five squares.

grid One or more sets of equidistant parallel segments or lines

grip A placement of the fingers of a robot hand on an object.

grip selection The problem of choosing the placement of the fingers of a robot hand on an object and the direction of the force to be applied by each finger.

gross planning The aspect of planning a robot's path that takes into account fixed aspects of the robot's environment.

ground speed The speed of an object relative to the ground.

group A set G of elements together with a BINARY OPERATION $*$ satisfying the following four axioms:

Closure: if a and b are in G, so is $a * b$;

Associativity: $(a * b) * c = a * (b * c)$;

Existence of an identity: there is e in G such that $e * a = a * e = a$;

Existence of inverses: for every a in G, there exists an element a^{-1} in G such that $a * a^{-1} = a^{-1} * a = e$.

group homomorphism *See* HOMOMORPHISM.

group of a knot A GROUP whose elements are used to label the arcs of a KNOT DIAGRAM so that specific algebraic relations hold at each crossing.

gyroelongated square dipyramid A DELTAHEDRON with 16 faces and 10 vertices.

half-line *See* RAY.

half-open interval An interval of real numbers that includes one endpoint but not the other. The notation $[a, b)$ means the half-open interval including a but not b and $(a, b]$ means the half-open interval including b but not a.

half-plane All the points on one side of a line in a plane. The line, which is not part of the half-plane, is called the edge of the half plane.

half-space All the points on one side of a plane in three-dimensional space.

half-turn A rotation of 180° about a given point.

Halmos symbol A small square, open or filled in, used to denote the end of a proof.

Hamiltonian circuit A circuit of edges in a GRAPH that is incident to every vertex once and only once.

Hamiltonian path A path of edges in a GRAPH that is incident to every vertex once and only once.

Hamming distance The distance between two points in an n-dimensional LATTICE, defined to be the number of places in which the points have different coordinates.

Hamming sphere All LATTICE POINTS within a given radius of a given point, with respect to the HAMMING DISTANCE.

hand *See* ROBOT ARM.

handle A cylinder attached to a surface by gluing each end of the cylinder to the boundary of a circular hole in the surface.

handle body A sphere with a finite number of HANDLES attached to it.

harmonic analyzer A mechanism for determining the FOURIER COEFFICIENTS of a given periodic function from its graph.

harmonic conjugate If H(AB, CD) is a HARMONIC SET, A and B are harmonic conjugates of each other with respect to C and D.

harmonic division Division of the segment AB by points C and D such that the CROSS RATIO (AB, CD) is -1.

harmonic homology A PERSPECTIVE COLLINEATION that is an involution, so termed because the harmonic conjugate of its center with respect to corresponding points is its axis.

harmonic mean The harmonic mean of the numbers a and b is $\dfrac{2ab}{a+b}$.

Handle body

harmonic net For three distinct collinear points, find the HARMONIC CONJUGATE of each with respect to the other two; then find the harmonic conjugate of any one of these points with respect to any two of them; continue in this way. The result will be a harmonic net.

harmonic pencil *See* HARMONIC SET OF POINTS.

harmonic range *See* HARMONIC SET OF LINES.

harmonic sequence A sequence of numbers in which the second of any three consecutive terms of the sequence is the HARMONIC MEAN of the first and third terms.

harmonic set of lines A set of four concurrent lines whose CROSS RATIO is −1. Equivalently, a harmonic set of lines consists of two diagonals of a complete quadrilateral and the two lines connecting the point of intersection of these diagonals with the two vertices of the quadrilateral that lie on the third diagonal.

harmonic set of points The QUADRANGULAR SET of points on a line passing through two diagonal points of a COMPLETE QUADRANGLE. Since each diagonal point is the intersection of a pair of sides of the quadrangle, a harmonic range will contain exactly four points. A harmonic set is denoted H(AB, CD) where one pair of points, either A and B or C and D, are the two diagonal points and the other pair are the other two points.

harmonically related Each point of a HARMONIC NET is harmonically related to the three points defining the net.

Hart's crossed parallelogram A mechanism consisting of four rods used to draw a straight line.

Hauptvermutung The main conjecture of COMBINATORIAL TOPOLOGY. It claims that every MANIFOLD is HOMEOMORPHIC to a TRIANGULATED MANIFOLD. It was shown to be false in the 1960s.

Hausdorff dimension *See* SELF-SIMILAR DIMENSION.

Hausdorff space A TOPOLOGICAL SPACE in which for two distinct points there are two disjoint OPEN SETS, each containing one of the points.

haversed sine One-half the VERSED SINE.

Hawaiian earring The union of a nested sequence of mutually tangent circles with diameters that tend to zero.

head The initial point of a vector.

head wind A wind blowing opposite to the direction of the course of a ship or plane.

heading The direction of the course of a ship, plane, or other vessel.

hectogon A polygon with 100 sides.

Heesch number For a closed plane figure, the maximum number of times the figure can be completely surrounded by copies of itself. For example, the Heesch number of a square is infinity since a square is surrounded by infinitely many copies of itself in the regular tiling by squares.

height The perpendicular distance from the base of a polygon or polyhedron to an opposite vertex or base.

helical polygon A polygon whose sides are chords of a helix.

helicoid A RULED SURFACE whose boundary is a helix. It can be generated by a segment perpendicular to a line moving with constant velocity along the line and constant angular velocity around the line.

helix A curve on the surface of a cylinder traced by a point rotating about the axis of the cylinder at constant speed while simultaneously moving in a direction parallel to the axis of the cylinder at a constant speed. Its curvature and torsion have a constant ratio.

hemisphere Half of a sphere.

hendacagon A polygon with 11 sides.

hendecahedron A polyhedron with 11 faces.

heptacontagon A polygon with 70 sides.

heptadecagon A polygon with 17 sides.

heptagon A polygon with seven sides.

heptahedron A polyhedron with seven faces. The heptahedron with three square faces, four equilateral triangular faces, 12 edges, and six vertices is a one-sided nonorientable surface topologically equivalent to the Roman surface.

heptiamond A polygon formed by joining seven equilateral triangles along their sides. There are 24 different heptiamonds.

hermit point *See* ISOLATED POINT.

Heronian triangle A triangle whose area is an integer and all of whose sides have integer length. In some contexts, a Heronian triangle has rational area and sides with rational length.

hexacaidecadeltahedron A DELTAHEDRON with 16 faces.

hexacontagon A polygon with 60 sides.

hexacontahedron A polyhedron with 60 faces.

hexadecagon A polygon with 16 sides.

hexaflexagon A hexagon folded from a flat strip of paper that can be "flexed" or turned inside out.

hexafoil A MULTIFOIL drawn outside a regular hexagon.

hexagon A polygon with six sides.

hexagonal close packing A PACKING of three-dimensional space by tangent spheres. The centers of the 12 spheres tangent to each sphere in the hexagonal close packing are the vertices of a TRIANGULAR ORTHOBICUPOLA.

hexagram A COMPOUND POLYGON consisting of two equilateral triangles; it is often called the Star of David.

hexahedron A polyhedron with six faces.

hexakaidecahedron A polyhedron with 16 faces.

hexakisicosahedron A polyhedron with 120 triangular faces. It is the DUAL of the great rhombicosidodecahedron.

hexakisoctahedron A polyhedron with 48 triangular faces. It is the DUAL of the great rhombicuboctahedron.

hexiamond A figure made of six congruent equilateral triangles joined along their sides. There are 12 different hexiamonds.

hexomino A figure made of six congruent squares joined along their sides. There are 35 different hexominos.

hidden edge An edge of a polyhedron that is not seen from a POINT OF PERSPECTIVITY. Determining the hidden edges of a polyhedron is an important problem in computer graphics.

higher-dimensional Having dimension three or higher.

Hilbert cube The unit cube in HILBERT SPACE; it is infinite-dimensional.

Hilbert space An infinite-dimensional space with an INNER PRODUCT.

hippopede A curve shaped like an elongated figure-eight, the hippopede is the intersection of a sphere with a thin cylinder that is internally tangent to the sphere. It was used by EUDOXUS to describe the apparent motion of the planet Jupiter.

HL   *See* HYPOTENUSE-LEG.

holonomy For a closed curve on a surface, the measure of the angle between a vector and its PARALLEL TRANSPORT around the curve. The holonomy of a closed planar curve is 0.

holyhedron A polyhedron with one or more holes in each face. A holyhedron cannot be convex.

homeogonal tiling A tiling in which any two vertices are related by a HOMEOMORPHISM of the plane that maps the tiling to itself.

homeohedral tiling A tiling in which any two tiles are related by a HOMEOMORPHISM of the plane that maps the tiling to itself.

homeomorphic Related by a HOMEOMORPHISM.

homeomorphism A CONTINUOUS function from one TOPOLOGICAL SPACE to another that has a continuous inverse. It preserves topological properties.

homeotoxal tiling A tiling in which any two edges are related by a HOMEOMORPHISM of the plane that maps the tiling to itself.

HOMFLY polynomial A polynomial determined from the sequence of CROSSINGS in a KNOT or LINK. It is a KNOT INVARIANT.

homogeneous Describing an equation or expression with the property that multiplying each instance of a variable by a constant will not change the value of the equation or expression. For example, $\frac{xy}{z^2} + \frac{x+y}{y}$ is a homogeneous expression because it is the same as $\frac{(ax)(ay)}{(az)^2} + \frac{ax+ay}{ay}$.

homogeneous coordinates A system of real-valued coordinates used in PROJECTIVE and AFFINE GEOMETRY. Each point in the PROJECTIVE PLANE is represented by coordinates *(x, y, z)* such that not all coordinates are zero and *(x, y, z)* and *(ax, ay, az)* with *a* nonzero represent the same point. Points of the form $(0, 0, z)$ are ideal points. Homogeneous coordinates are as powerful in projective geometry as Cartesian coordinates are in Euclidean geometry and are used extensively in the computations associated with computer graphics.

homogeneous polynomial A polynomial in which every term has the same degree. For example, $x^2 + xy + yz$ is homogeneous whereas $x^2y + x^2$ is not.

homogeneous tiling *See* SEMIREGULAR TILING.

homographic Two RANGES or PENCILS are homographic if there is a one-to-one correspondence between them that preserves CROSS RATIO.

homographic transformation *See* MÖBIUS TRANSFORMATION.

homography (1) A product of an even number of INVERSIONS. (2) On the complex plane, a homography is a MÖBIUS TRANSFORMATION of the form $f(x) = (az + b)/(cz + d)$ where $ad - bc = 1$.

homological algebra The study of the properties of homology group from an algebraic perspective.

homologous elements Two elements that correspond to each other under a one-to-one correspondence.

homology (1) A PERSPECTIVE COLLINEATION in which the center does not lie on the axis. (2) *See* DILATIVE ROTATION.

homology group A GROUP belonging to a HOMOLOGY THEORY.

homology theory An assignment of GROUPS, indexed by DIMENSION, to a TOPOLOGICAL SPACE or MANIFOLD. A CONTINUOUS function from one space to another induces a HOMOMORPHISM between homology groups with the same index.

homomorphism A function from one GROUP to another that preserves the group operation. Thus, if h is a homormophism, $h(a * b) = h(a) * h(b)$.

homothecey *See* DILATION.

homothetic Related by a DILATION.

homothetic transformation *See* DILATION.

homothety *See* DILATION.

homotopy A transforming of one path, loop, or space to another through a continuous stretching, shrinking, or deforming. A homotopy might not be a homeomorphism. For example, there is a homotopy from a cylinder to a circle.

homotopy group For a topological space, the group of loops, all having the same starting and ending points. The group operation is defined by traveling along one loop and then the other. Two loops are equivalent if one can be transformed to the other by a homotopy.

homotopy theory The study of homotopies, HOMOTOPY GROUPS, and their properties.

honeycomb A solid tessellation; a filling of three-dimensional space with polyhedra such that a face of one polyhedron is the face of exactly one other polyhedron.

Hopf link A simple LINK consisting of two circles.

horizon The BLOCKING SEGMENTS furthest from a VIEWPOINT on a TERRAIN.

horizon line In a perspective drawing, the line representing the juncture of the land with the sky. It is the projection onto the picture plane of the horizontal plane passing through the eye of the viewer.

horizontal line test A test used to determine whether a function has an INVERSE using the graph of the function. If every horizontal line cuts the graph in at most one point, then the function has an inverse; otherwise, it does not.

horocycle In HYPERBOLIC GEOMETRY, a unit circle whose center is a POINT AT INFINITY. Any two points on a horocycle are related by PARALLEL DISPLACEMENT with respect to the line at infinity.

horopter A curve that is the intersection of a cylinder and a HYPERBOLIC PARABOLOID.

horosphere In HYPERBOLIC GEOMETRY, a sphere whose center is a POINT AT INFINITY. Any two points on a horosphere are related by PARALLEL DISPLACEMENT with respect to the plane at infinity. The geometry on a horosphere satisfies EUCLIDEAN AXIOMS.

horse fetter *See* HIPPOPEDE.

horseshoe map A TRANSFORMATION from the plane to itself, invented by STEPHEN SMALE, that maps a given square to a horseshoe region overlapping the position of the original square. Under repeated iterations, this map exhibits chaotic behavior.

hosohedron A MAP on a sphere with two vertices, all of whose faces are DIGONS.

hyperbola The intersection of a double cone with a plane that intersects both NAPPES of the cone. Equivalently, the locus of all points whose distances from two points, called the foci of the hyperbola, have a constant difference. *See also* CONIC SECTION.

hyperbolic axiom The axiom that states that at least two lines can be drawn through a given point that is not on a given line that will not intersect the given line. This is used instead of Euclid's fifth postulate to define HYPERBOLIC GEOMETRY.

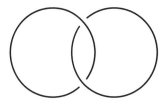

Hopf link

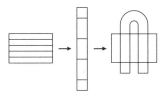

Horseshoe map

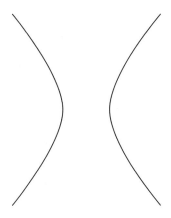

Hyperbola

hyperbolic cosine The function of a complex variable z defined by

$$\cosh z = \frac{e^x + e^{-x}}{2} = \cos(iz).$$

hyperbolic cylinder A cylinder whose base is a hyperbola; it has two sheets.

hyperbolic geometry A non-Euclidean geometry in which Euclid's parallel postulate is replaced by a postulate asserting that for a line and a point not on the line, there exists more than one line going through the point and parallel to the given line. Models of hyperbolic geometry include the PSEUDOSPHERE and the POINCARÉ DISC.

hyperbolic paraboloid A saddle-shaped ruled QUADRIC SURFACE given by the equation $x^2/a^2 - y^2/b^2 = z$.

hyperbolic plane A plane satisfying the axioms of HYPERBOLIC GEOMETRY.

hyperbolic point A point on a surface where the surface lies on both sides of the tangent plane.

hyperbolic projectivity *See* PROJECTIVITY.

hyperbolic rotation An EQUIAFFINITY of the form $f(x, y) = (ax, a^{-1} y)$ such that $a > 0$.

hyperbolic sine The function of a complex variable z defined by

$$\sinh z = \frac{e^x - e^{-x}}{2} = \sin(iz).$$

hyperbolic umbilic A HYPERBOLIC POINT that is also an UMBILIC.

hyperboloid The surface of revolution of a hyperbola about one of its two axes of symmetry. If the axis is the line connecting the foci, the hyperboloid has two sheets and if the axis is perpendicular to the line connecting the focus, the hyperboloid has one sheet.

hyperboloidal gears Gears shaped like two congruent HYPERBOLOIDS; they mesh along the GENERATORS and transform a circular motion around one axis to a circular motion around another axis.

hypercube A four-dimensional cube; it is a four-dimensional polytope with 16 vertices and eight cells, each a cube. In some contexts, a hypercube is a cube of dimension four or higher.

hypercycle *See* EQUIDISTANT CURVE.

hyperfixed A point is hyperfixed by a TRANSFORMATION if every HYPERPLANE through the point is mapped to itself by the transformation.

hyperparallel *See* ULTRAPARALLEL.

hyperplane A higher-dimensional analogy of the plane. It can be realized as the graph of a linear equation in four or more variables.

hyperplane at infinity A HYPERPLANE added to a Euclidean space that consists of IDEAL POINTS.

hypersphere A higher-dimensional sphere. It is the set of all points equidistant from a given point in a space of four or more dimensions.

hypocycloid The locus of a point fixed on the circumference of a circle that rolls on the inside of another circle without slipping.

hypotenuse In a right triangle, the side opposite the right angle.

hypotenuse-leg If the hypotenuse and leg of one triangle are proportional to the hypotenuse and leg of another, the two triangles are similar.

hypothesis (1) Something assumed in order to prove a given result. (2) The statement p in the implication "If p, then q."

hypotrochoid The locus of a point fixed on a ray from the center of a circle that rolls without slipping inside another circle.

i   The number whose square is -1. Thus $i^2 = -1$.

icosagon A polygon with 20 sides.

icosahedron A polyhedron with 20 congruent faces, each an equilateral triangle. Sometimes, any polyhedron with 20 faces.

icosidodecahedron A SEMIREGULAR POLYHEDRON with 20 equilateral triangular faces and 12 regular pentagonal faces.

icosioctahedron A polyhedron with 28 faces.

icosohedral group The GROUP of DIRECT SYMMETRIES of an icosahedron; it has 60 elements. It is isomorphic to the group of direct symmetries of the dodecahedron.

ideal line A line added to the Euclidean plane consisting of IDEAL POINTS.

ideal plane A plane added to three-dimensional Euclidean space consisting of IDEAL LINES.

ideal point A point added to each line in a family of parallel lines in a geometric space. Then, any two lines in the family will intersect at the ideal point.

identification space A TOPOLOGICAL SPACE formed from another space by gluing together or "identifying" certain points. For example, a circle

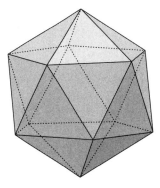

Icosahedron

is the identification space formed from an interval by identifying the two endpoints.

identity (1) For a BINARY OPERATION ∗, the identity is the element e with the property $a * e = e * a = a$ for each element a. For addition, the identity is 0 and for multiplication the identity is 1. (2) The function or transformation that takes each element to itself, often denoted I or id.

identity matrix The identity element for MATRIX MULTIPLICATION; it is a square matrix whose elements are 0 except on the MAIN DIAGONAL where they are 1. It is denoted I.

if . . . then The usual form of an IMPLICATION statement.

iff If and only if.

IFS *See* ITERATED FUNCTION SYSTEM.

image (1) The set of points which are the outputs of a function. The image of a function is a subset of the range of the function. (2) A two-dimensional picture.

image plane The plane on which an image is located.

image point A point on an image plane.

imaginary axis The vertical axis of the COMPLEX PLANE. The unit distance along the imaginary axis is equal to i.

imaginary number *See* COMPLEX NUMBER.

imaginary part The coefficient of i in a complex number. For the complex number $a + bi$, the imaginary part is b.

implication A statement of the form "If p, then q" where p and q are statements; it can be written in the form "$p \rightarrow q$". In this statement, p is a sufficient condition for q and q is a necessary condition for p.

in perspective Related by a PERSPECTIVITY.

incenter The center of an INCIRCLE or INSPHERE.

incidence A relationship based on containment or inclusion. Two geometric objects are incident if one of them contains the other. Thus, if point A is on line l, A and l are incident to each other.

incidence, angle of The angle between the trajectory or path of a particle or ray of light hitting a surface and the normal to the surface at that point.

incidence axiom An axiom that gives INCIDENCE RELATIONSHIPS among the objects of a geometric theory.

incidence geometry *See* PROJECTIVE GEOMETRY.

incidence matrix For a graph, a square matrix with one row and one column for each vertex in the graph; the entry in the matrix is 1 if there is an edge between the vertices corresponding to the row and the column of the entry and 0 otherwise.

incidence number The number of points contained in the intersection of two distinct geometric objects.

incidence relation A relation between two objects that shows containment of one object by the other. Thus a line and a point on the line are related by incidence. This relation is symmetric and transitive but not reflexive.

incidence structure A geometry consisting of point and lines, in which the axioms and theorems describe the INCIDENCE RELATIONSHIPS.

incidence table For a PROJECTIVE GEOMETRY, a table with one row for each point in the geometry and one column for each line in the geometry. The entry in the table is 1 if the point corresponding to the row of the entry lies on the line corresponding to the column of the entry and 0 otherwise.

incident Describing two objects such that one of them is contained in the other.

incircle *See* INSCRIBED CIRCLE.

inclination, angle of In general, the angle between two lines or two planes. In astronomy, the angle of inclination is the angle between the orbit of a planet and the ecliptic.

include To contain.

included angle The angle between two adjacent sides of a polygon.

included arc The arc of a circle in the interior of an angle.

included side The side between two adjacent angles of a polygon.

incommensurable Not having a common measure. For example, a segment of length 1 and a segment of length $\sqrt{2}$ are incommensurable.

independence The property of a collection of AXIOMS that no one of the axioms can be proved from the other axioms.

independent points Points in GENERAL POSITION.

independent statement A statement made using the terms of an AXIOMATIC SYSTEM that cannot be proved or disproved within the system.

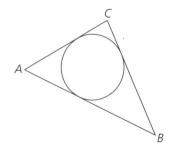

The incircle of triangle
ABC

independent variable A variable whose value may be freely chosen.

index of a point The WINDING NUMBER of a curve with respect to the point.

indirect isometry *See* OPPOSITE ISOMETRY.

indirect measurement Deriving a new measurement by geometric reasoning from given measurements.

indirect proof *See* PROOF BY CONTRADICTION.

induce To give rise to, usually by restriction. For example, a METRIC on a three-dimensional space induces a metric on any plane it contains.

induced tile group The GROUP of SYMMETRIES of a tiling that map a given tile to itself.

induction (1) A method of proving that a statement is true for all positive integers by first proving that the statement is true for the integer 1 and then proving that if the statement is true for some integer, it is true for the next larger integer. (2) The process of determining a general principle from specific examples.

inequality An expression showing that one quantity is greater than or less than another quantity. The symbols used in inequalities are the following: <, less than; ≤, less than or equal to; >, greater than; ≥, greater than or equal to.

infimum For a set of numbers, the largest number that is smaller than every element of the set.

infinitesimal motion *See* FLEX.

infinitesimally rigid Describing a FRAMEWORK or MECHANISM with the property that the only INFINITESIMAL MOTIONS are congruences of the whole framework or mechanism.

inflation Creation of a new tiling by DECOMPOSITION of a tiling followed by an EXPANSION. The tiles of the new tiling are the same size as the tiles of the original tiling.

inflection point A point on a curve where the CURVATURE changes sign.

initial condition The smallest value or values for the parameters belonging to a curve, DYNAMICAL SYSTEM, or other parameterized object.

initial point (1) The starting point of an ORIENTED curve. (2) The starting point of a vector.

initial side *See* DIRECTED ANGLE.

injection A function that is ONE-TO-ONE.

injective *See* ONE-TO-ONE.

inner product *See* DOT PRODUCT.

inner symmetry In physics, a LOCAL SYMMETRY of SPACETIME GEOMETRY.

inparameter A measure of the size of the sets in a collection of sets. It is a number u such that each set in the collection contains a ball of radius u.

inradius The radius of an incircle or insphere. *See* INSCRIBED CIRCLE and INSCRIBED SPHERE.

inscribed angle An angle whose vertex is on the circumference of a circle and whose legs are chords of the circle.

inscribed circle A circle that is tangent to every side of a polygon. The center of the inscribed circle of a triangle is the intersection of the bisectors of the angles of the triangle.

inscribed cone of a pyramid A cone that has the same vertex as a pyramid and whose base is a circle inscribed in the base of the pyramid.

inscribed polygon A polygon all of whose vertices are on the circumference of a circle.

inscribed sphere A sphere that is tangent to all the faces of a polyhedron.

insphere *See* INSCRIBED SPHERE.

instability domain The values of the PARAMETERS of a DYNAMICAL SYSTEM for which the EQUILIBRIUM STATE is UNSTABLE.

instantaneous center of rotation The intersection of the lines perpendicular to the velocity vectors at any two points on a moving rigid body. This point is momentarily at rest.

integer A whole number such as –4, –1, 0, 2, or 5.

integer lattice *See* STANDARD LATTICE.

integral In CALCULUS, an expression that represents the sum of many small pieces. Integrals can be used to compute areas, volumes, and arc lengths.

integral geometry *See* GEOMETRIC PROBABILITY.

intercept To cut off or to intersect. Also, a point or points of intersection.

intercepted arc The portion of an curve or circle between two points; the intercepted arc of an angle is the portion of the curve or circle in the interior of the angle.

interior The finite region enclosed by a closed curve or surface.

interior angle An angle that is formed by two adjacent sides of a polygon and whose interior contains all or part of the interior of the polygon.

interior angles The two angles that are between two lines and are on the same side of a TRANSVERSAL crossing the lines. If the lines are parallel, the interior angles are SUPPLEMENTARY.

interior point A point of a set that is an element of an OPEN SET that is completely contained in the set.

intermediacy The property of being between two points or other objects.

internal Inside or interior to.

internal division of a line The division of a line segment by a point between the endpoints. If P is between A and B, the segment AB is divided internally by P with ratio AP/PB.

interplanar spacing The distance from the origin to the closest of a family of parallel RATIONAL PLANES that contain all the points of a LATTICE.

interpolation The use of a straight line or other simple curve to approximate the points between two given points on a more complex curve or surface.

interpretation An assignment of a meaning to each of the UNDEFINED TERMS of an AXIOMATIC SYSTEM.

intersect To have at least one point in common. Two sets are said to intersect if their intersection is nonempty. *See* INTERSECTION.

intersection The set that contains all elements contained in every set in a given collection of sets. The notation $S \cap T$ means the intersection of the set S with the set T.

interval The set of all points between two distinct points on the number line. The interval between a and b is denoted (a, b) and does not include a or b. *See also* CLOSED INTERVAL.

intrinsic Related to the internal structure of a geometric shape without reference to the space containing it.

intrinsic curvature CURVATURE with respect to measurements taken on a surface rather than with respect to the space the surface is EMBEDDED in.

intrinsic distance The distance between two points on a surface. It is the length of the shortest curve on the surface that connects the two points. For example, the intrinsic distance between two ANTIPODAL points on a sphere or radius 1 is π, although the EXTRINSIC DISTANCE between the two points is 2.

invariant Fixed or unchanged by a given TRANSFORMATION or GROUP of transformations.

inverse cosecant The inverse function of the COSECANT, defined so that its range is $(-90°, 90°)$. Thus, since csc $(30°) = 2$, $\csc^{-1}(2) = 30°$.

inverse cosine The inverse function of the COSINE, defined so that its range is $[0°, 180°]$. Thus, since cos $(60°) = 1/2$, $\cos^{-1}(1/2) = 30°$.

inverse cotangent The inverse function of the COTANGENT, defined so that its range is $(0°, 180°)$. Thus, since cot $(45°) = 1/\sqrt{2}$, $\cot^{-1}(1/\sqrt{2}) = 45°$.

inverse function A function that reverses the action of a given function. For example, if $f(x) = 2x$, its inverse function, denoted f^{-1}, is the function $f^{-1}(x) = 1/2\ x$. A function composed with its inverse is the identity function.

inverse kinematic problem Calculation of the angles between the links of a ROBOT ARM that correspond to a given position and orientation of the robot hand.

inverse locus The set of points that are the images under INVERSION of the points of a given locus.

inverse of a knot The KNOT formed by taking the REVERSE of the MIRROR IMAGE of the given knot.

inverse of a statement *See* INVERSE STATEMENT.

inverse point The IMAGE of a POINT under INVERSION.

inverse secant The inverse function of the SECANT, defined so that its range is $[0°, 180°]$. Thus, since sec $(60°) = 2$, $\sec^{-1}(2) = 60°$.

inverse sine The inverse function of the SINE, defined so that its range is $[-90°, 90°]$. Thus, since sin $(30°) = 1/2$, $\sin^{-1}(1/2) = 30°$.

inverse statement The inverse of the implication $p \rightarrow q$ is the implication $\sim p \rightarrow \sim q$.

inverse tangent The inverse function of the TANGENT, defined so that its range is $(-90°, 90°)$. Thus, since tan $(45°) = 1/\sqrt{2}$, $\tan^{-1}(1/\sqrt{2}) = 45°$.

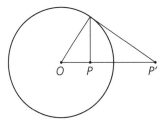

Point *P′* is the inverse of point *P* with respect to the circle with center *O*

inversion Inversion is a function from the plane to itself that leaves a given circle, called the circle of inversion, fixed and maps the interior of the circle of inversion (except for its center C) to the exterior of the circle and vice versa. Inversion maps a point P to the point P' on the ray CP such that $(CP)(CP') = r^2$, where r is the radius of the circle of inversion. The domain of an inversion is all points in the plane except for the center of the circle of inversion and is equal to the range. Inversion in three-dimensional space with respect to a sphere is defined similarly.

inversion through a point *See* POINT SYMMETRY.

inversive distance A measurement of the distance between two nonintersecting circles in the INVERSIVE PLANE. The inversive distance between two circles is the absolute value of the NATURAL LOGARITHM of the ratio of the radii of two concentric circles that are the inversive images of the two circles.

inversive geometry *See* CIRCLE GEOMETRY.

inversive plane The Euclidean plane with a point that has been added at infinity to make inversion a one-to-one correspondence. The added point, called an ideal point, is contained in every line in the inversive plane and is the image of the center of inversion for every INVERSION.

involutory transformation *See* INVOLUTION.

involute An involute of a curve can be obtained by unwinding a string that is attached to one point on the curve and stretched along the curve; as the string is unwound while keeping it tight, the free end traces out the involute of a curve. An involute of a curve is a curve whose evolute is the given curve. One curve can have more than one involute. For example, the evolute of a circle is a spiral.

involution A transformation that, when composed with itself, gives the IDENTITY TRANSFORMATION.

involutoric transformation *See* INVOLUTION.

irrational A real number such as $\sqrt{2}$ or π that cannot be written as a ratio of two integers. Irrationals can be represented by nonrepeating, nonterminating decimals.

irrotational Not having ROTATIONAL SYMMETRY.

isocenter A point such that every angle with vertex at that point is transformed to a congruent angle by a PERSPECTIVITY.

isocline For a vector field, a curve that passes through points whose assigned vectors all have the same direction.

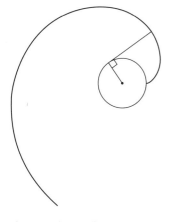

The involute of a circle

isodynamic point The ISOGONAL CONJUGATE of an ISOGONIC CENTER of a triangle.

isogonal conjugate If two lines are ISOGONAL LINES, one is the isogonal conjugate of the other.

isogonal conjugate points Construct ISOGONAL LINES at each vertex of a triangle. If three of these lines, one at each vertex, are concurrent, so are the other three lines. The two points of concurrency are isogonal conjugate points.

isogonal lines Two rays in the interior of an angle that make congruent angles with the angle bisector.

isogonal mapping *See* CONFORMAL MAPPING.

isogonal tiling A tiling in which all vertices belong to the same TRANSITIVITY CLASS. If there are *k* transitivity classes of vertices, the tiling is *k*-isogonal.

isogonal vertices Vertices of a tiling that have congruent VERTEX FIGURES.

isogonic center For a given triangle, construct equilateral triangles on each side, either all internal to the given triangle or all external. Connect each new vertex to the opposite vertex of the original triangle. The point of concurrency of these three lines is an isogonic center of the triangle. Each triangle has two isogonic centers, one for the internal triangles and one for the external triangles.

isohedral Having congruent faces.

isohedral tiling A MONOHEDRAL TILING in which all the tiles belong to the same TRANSITIVITY CLASS.

isolated point In a set of points, a point with a NEIGHBORHOOD that contains no other points of the set.

isoline A line, other than the AXIS OF PERSPECTIVITY, that is transformed ISOMETRICALLY by a PERSPECTIVITY.

isometric Distance-preserving.

isometry A TRANSFORMATION that preserves distances.

isomorphism A one-to-one correspondence that preserves structure. For example, an isomorphism of PROJECTIVE SPACES preserves the INCIDENCE RELATION.

isonemal fabric A pattern used to model woven fabric; it consists of infinitely long layered strips.

isoperimetric Having the same perimeter. There are two main types of problems called isoperimetric problems: finding, of all figures with a given property, that one which has the least perimeter or surface area and finding, of all figures with a given perimeter or surface area, that one which has the greatest area or volume.

isosceles trapezoid A TRAPEZOID whose two legs are congruent. In an isosceles trapezoid, angles adjacent to the same base are congruent.

isosceles triangle A triangle in which two sides are congruent.

isotomic conjugate points Choose a pair of ISOTOMIC POINTS on each side of a triangle and construct three lines, each connecting a different vertex of the triangle to one of the isotomic points on the opposite side. If these lines are concurrent, so are the lines connecting the vertices to the other isotomic points. Two such points of concurrence are isotomic conjugate points.

isotomic points Two points on a segment form a pair of isotomic points if they are equidistant from the midpoint of the segment and are on opposite sides of the midpoint.

isotopic The isotopic of a curve is the locus of the point of intersection of two lines tangent to the curve that intersect at a constant angle.

isotopic tilings Tiling with the property that one can be continuously DEFORMED into the other.

isotoxal tiling A tiling in which each edge is EQUIVALENT to every other edge.

isozonohedron A ZONOHEDRON whose faces are congruent to one another.

iterated function system A finite set of functions that are CONTRACTIONS. Iterated function systems can be used to define DYNAMICAL SYSTEMS and create fractals.

jack One of the seven different vertex neighborhoods that can occur in a PENROSE TILING.

JavaSketchpad An extension of Geometer's Sketchpad that creates interactive sketches for webpages.

jitterbug A physical model of a CUBOCTAHEDRON with flexible joints that can be twisted into an icosahedron, then an octahedron, and finally a tetrahedron. It was invented by BUCKMINSTER FULLER.

Johnson solid A convex polyhedron with regular faces and congruent edges that is not a PLATONIC SOLID, an ARCHIMEDEAN SOLID, a PRISM, or an ANTIPRISM. There are 92 different Johnson solids.

join The join of two points is the line containing them.

joint   *See* ROBOT ARM.

Jones polynomial   A polynomial determined from the sequence of crossings in a KNOT or LINK. It is a KNOT INVARIANT.

Jordan curve   A SIMPLE CLOSED CURVE. Examples are a circle and a TREFOIL KNOT.

Jordan loop   *See* JORDAN CURVE.

Julia fractal   A JULIA SET that is a fractal.

Julia set   The BOUNDARY of the PRISONER SET of a DYNAMICAL SYSTEM.

***k*-isogonal**   *See* ISOGONAL TILING.

***k*-isohedral tiling**   A tiling in which there are *k* TRANSITIVITY CLASSES of tiles.

***k*-rep tile**   A REPTILE that can be cut into exactly *k* congruent pieces. *See also* REP-*k* tile.

kaleidoscope   A device for creating DIHEDRAL designs made by joining mirrors along an edge. If the angle between the mirrors is $180°/n$, for *n* an integer, one will see a design with *n*-fold symmetry when an object is placed between the mirrors.

kaleidoscopic symmetry   *See* DIHEDRAL SYMMETRY.

Kauffman polynomial   A polynomial determined from the sequence of CROSSINGS in a KNOT or LINK. It is a KNOT INVARIANT.

Kepler conjecture   The conjecture by JOHANNES KEPLER that the densest PACKING by spheres is the CUBIC CLOSE PACKING. It was proved to be true in 1998.

Kepler-Poinsot solids   The four regular nonconvex star polyhedra: the small stellated dodecahedron, the great stellated dodecahedron, the great dodecahedron, and the great icosahedron.

keratoid cusp   *See* CUSP OF THE FIRST KIND.

kernel   The set of all VECTORS mapped to the zero vector by a LINEAR TRANSFORMATION. It is a SUBSPACE of the domain of the linear transformation.

king   One of the seven different VERTEX NEIGHBORHOODS that can occur in a PENROSE TILING.

Kirkman point   A point of intersection of three PASCAL LINES. Any six points inscribed in a conic section determine 60 Kirkman points.

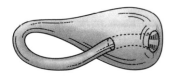

Klein bottle

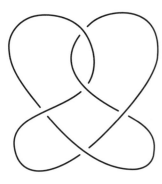

A knot diagram with five crossings

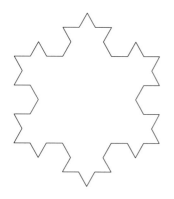

Koch snowflake

kissing number The largest number of congruent HYPERSPHERES in n dimensions that can touch a congruent hypersphere so that any two of the hyperspheres are tangent or disjoint. For dimension 2, the kissing number is 6 since 6 nonintersecting circles can be placed tangent to one circle.

kite A CONVEX quadrilateral with two pairs of adjacent congruent sides. Also, a PENROSE TILE shaped like a kite.

Klein bottle A NONORIENTABLE closed surface that cannot be EMBEDDED in three-dimensional space. It can be formed by attaching two MÖBIUS BANDS along their edges.

Klein four-group The GROUP of SYMMETRIES of a rectangle. It has four elements: the identity, a half-turn, and two reflections.

Klein's surface *See* KLEIN BOTTLE.

knot (1) A non-self-intersecting CLOSED CURVE in three-dimensional space. A knot can be drawn on paper with only a finite number of crossings. Two knots are EQUIVALENT if one can be deformed into the other. (2) One nautical mile per hour.

knot complement The set of all points in three-dimensional space that do not belong to a specific KNOT.

knot diagram A planar diagram representing the PROJECTION of a KNOT onto the plane. Intersections in the projection are replaced by breaks, called crossings. Two knot diagrams are equivalent if they both represent the same knot.

knot projection *See* KNOT DIAGRAM.

Koch snowflake A fractal curve in which each edge of an equilateral triangle is replaced by four congruent edges ⌐⌐ pointing outward and then this process is repeated infinitely.

Kronecker delta A function of two variables that is 1 if they are equal and 0 otherwise, symbolized by δ. For example, $\delta(1, 1) = 1$ and $\delta(1, 2) = 0$. Sometimes, $\delta(i, j)$ is written δ_{ij}.

Lagrange identity For VECTORS **u**, **v**, **w**, and **x,** the SCALAR equation $(\mathbf{u} \times \mathbf{v}) \cdot (\mathbf{w} \times \mathbf{x}) = (\mathbf{u} \cdot \mathbf{w})(\mathbf{v} \cdot \mathbf{x}) - (\mathbf{u} \cdot \mathbf{x})(\mathbf{v} \cdot \mathbf{w})$.

Lagrange polynomial The unique polynomial of degree n that passes through $n + 1$ given points in a plane.

lakes of Wada Three regions ("lakes") that are so convoluted that a NEIGHBORHOOD of a point contained in any one of the regions will include points from the other two.

Lambert conformal projection A CONFORMAL PROJECTION of the Earth's surface onto a cone whose axis is the polar axis and which is tangent or secant to a particular region of interest.

Lambert quadrilateral A quadrilateral with three right angles. It is used in the study of HYPERBOLIC GEOMETRY.

lamina A thin surface of uniform DENSITY.

Laplacian The DIVERGENCE of the GRADIENT of a SCALAR FIELD.

latent root *See* EIGENVALUE.

lateral face A face of a polyhedron other than a base.

latitude A circle on the surface of the Earth that is the intersection of the Earth with a plane perpendicular to the north-south axis.

lattice An infinite regular array of points in any dimensional space. The symmetries of a lattice of dimension n include TRANSLATIONS in n different directions.

lattice complex *See* DOT PATTERN.

lattice group The GROUP of SYMMETRIES of a LATTICE; it is generated by nonparallel TRANSLATIONS.

lattice metric *See* TAXICAB METRIC.

lattice of a periodic tiling The LATTICE formed from a point in a PERIODIC TILING and all its translates under SYMMETRIES of the tiling.

lattice path A sequence of LATTICE POINTS with the property that consecutive points are one unit away from each other.

lattice point A point contained in a LATTICE.

lattice-point-free path A path in the plane of a LATTICE that contains no LATTICE POINTS.

lattice polygon A polygon whose vertices are points of a LATTICE.

lattice system The lines whose intersections form a LATTICE.

lattice unit A parallelogram or parallelepiped whose vertices are LATTICE POINTS and whose sides or edges are parallel to the translation vectors that generate the lattice.

latus rectum The chord of a conic section passing through the focus and perpendicular to the axis of the conic section. It is thus parallel to the directrix of the conic section.

Laurent polynomial A polynomial determined from the sequence of CROSSINGS in a KNOT or LINK. It is a KNOT INVARIANT equivalent to the Alexander polynomial.

Laves tiling A MONOHEDRAL TILING all of whose vertices are regular.

law of excluded middle The logical principle that says a statement is either true or false.

leaf *See* FOLIATION.

Leech lattice The LATTICE corresponding to a dense PACKING of spheres in 24-dimensional space.

left-handed Having the same ORIENTATION as the thumb and first two fingers of the left hand held open with the thumb up, the index finger extended, and the middle finger bent.

leg A side of a polygon. In an isosceles triangle, a leg is one of the two equal sides; in a right triangle, a leg is one of the two sides adjacent to the right angle; in a trapezoid, a leg is one of the two nonparallel sides.

lemma A proposition or theorem, especially one that is used in the proof of a more significant theorem.

lemniscate of Bernoulli The locus of points the product of whose distances to two points at a distance a apart is a^2. It is given by the polar equation $r^2 = a^2 \cos 2\theta$ for a constant.

Lemoine axis For a each vertex of a given triangle, find the point of intersection of a line tangent to the circumcircle at the vertex with the extension of the side of the triangle opposite the vertex. These three points are collinear and lie on the Lemoine axis.

Lemoine circle *See* FIRST LEMOINE CIRCLE and SECOND LEMOINE CIRCLE.

Lemoine point *See* SYMMEDIAN POINT.

length The distance between the two endpoints of a segment. *See also* ARC LENGTH.

lens A CONVEX figure bounded by two congruent circular arcs.

level curve A level curve contains all points in the domain of a SCALAR FUNCTION of two variables that have the same image.

level line *See* LEVEL CURVE.

level surface A level surface contains all points in the domain of a SCALAR FUNCTION of three variables that have the same image.

Levi-Civita parallelism *See* PARALLEL TRANSPORT.

levo An ENANTIOMER that rotates polarized light to the left.

Lie group A MANIFOLD that is also a GROUP. Examples are the circle, the real number line, and the set of all ISOMETRIES of three-dimensional space.

ligancy *See* KISSING NUMBER.

light cone The set of points in a SPACETIME GEOMETRY of norm 0. Each point on the light cone corresponds to an event that is simultaneous with the origin.

limaçon of Pascal The CONCHOID determined by a circle and a point on the circle. It is given in polar coordinates by the equation $r = 2a \cos \theta + k$ for a and k constant.

limit *See* SEQUENCE.

limit point A point whose NEIGHBORHOODS all contain infinitely many points of a given set.

limiting condition A condition in the statement of a theorem that restricts the properties of the givens.

limiting curve In HYPERBOLIC GEOMETRY, the set of all points on a PENCIL of parallels that correspond to a given point on one of the parallels.

limiting cycle *See* HOROCYCLE.

limiting points The two points common to all circles orthogonal to each of two nonintersecting circles.

line A set of points that is straight, has infinite extent in two opposite directions, and has length but no breadth. *Line* is usually an undefined term in a geometry and is used interchangeably with *straight line*. Following EUCLID, lines are usually represented by a lowercase letter or by two points on the line.

line at infinity *See* IDEAL LINE.

Line Axiom The statement that any two distinct points determine a unique line.

line conic A line conic is the reformulation of the concept of a CONIC SECTION into the language of PROJECTIVE GEOMETRY. A line conic is the set of lines that join corresponding points in two ranges that are related by a PROJECTIVITY but not a PERSPECTIVITY.

line of centers The line passing through the centers of the circles in a given collection.

line of curvature A curve on a surface that is tangent to one of the PRINCIPAL DIRECTIONS of curvature of each of its points.

line of perspectivity If two polygons are PERSPECTIVE FROM A LINE, the line is the line of perspectivity for the polygons.

line of sight A ray starting from a viewer's eye or from a viewpoint.

line symmetry *See* BILATERAL SYMMETRY.

linear algebra The algebra and geometry of VECTOR SPACES and MATRICES.

linear combination A sum of SCALAR MULTIPLES of one or more VECTORS.

linear element An infinitesimally small line approximating a curve at a point.

linear equation An equation whose graph is a line or plane. Thus, a linear equation is a polynomial equation of degree 1.

linear fractional map *See* MÖBIUS TRANSFORMATION.

linear mapping *See* LINEAR TRANSFORMATION.

linear programming The process of finding a maximum or minimum value for a function defined on a CONVEX BODY.

linear set A set with the property that for every pair of points in the set, every point on the line joining them is also in the set.

linear space A geometric space in which any two distinct points determine a line.

linear subspace A subset of a LINEAR SPACE that is itself a linear space.

linear transformation A function T from one VECTOR SPACE to another that preserves VECTOR ADDITION and SCALAR MULTIPLICATION: $T(\mathbf{v} + \mathbf{w}) = T(\mathbf{v}) + T(\mathbf{w})$ and $T(a\mathbf{v}) = a\,T(\mathbf{w})$ for for vectors $\mathbf{v}$ and $\mathbf{w}$ in the range and scalar a. A linear transformation can be represented by a MATRIX.

linearly dependent Describing a set of VECTORS such that the zero vector is a LINEAR COMBINATION of the vectors with not all scalars equal to 0.

linearly independent Describing a set of VECTORS such that the zero vector is a LINEAR COMBINATION of the vector only when all scalars are equal to 0.

linearly separable Describing two sets in the plane such that there is a line separating them; the points of one set are in one half-plane determined by the line and the points of the other set are in the other half-plane.

Lineland A line on which imaginary one-dimensional creatures live, as described by British clergyman E. A. ABBOTT in the satire *Flatland*.

linewise fixed A point is linewise fixed by a TRANSFORMATION if every line through the point is mapped to itself by the transformation.

link (1) Two or more disjoint KNOTS. (2) *See* ROBOT ARM.

linkage A mechanical system of rigid rods, connected to each other or to fixed points at joints which are free to turn.

linking number The linking number is a way to measure how intertwined are the knots making up a LINK. For a link consisting of two knots, the linking number is one half of the difference of the number of POSITIVE CROSSINGS between the two knots minus the number of NEGATIVE CROSSINGS between the two knots. It is a KNOT INVARIANT.

list color To color the VERTICES of a GRAPH, where each vertex has a list of acceptable colors, so that adjacent vertices have different colors.

Listing knot *See* FIGURE-EIGHT KNOT.

lituus The spiral given by the polar equation $r^2 \theta = a^2$. It has the ray $\theta = 0$ as asymptote and is often used as the volute at the top of an Ionic column.

loaded wheelbarrow *See* PENROSE'S WHEELBARROW.

lobster A HEXIAMOND shaped like a lobster. It is a REPTILE.

local Valid only in a small part of a region.

local coordinates A continuous one-to-one correspondence of an open disc in the plane to a region on a surface. Every point in the region is given the coordinates belonging to its pre-image in the plane.

locally finite Describing an infinite family of sets with the property that every point has an open NEIGHBORHOOD that intersects only finitely many sets in the family.

loci, method of A method of solving a construction problem by finding the intersection of appropriate loci.

locus A set of points, especially the set of all points satisfying a certain property or geometrical condition.

logarithm An exponent with respect to a given base. For example, the logarithm for base 10 of 1,000 is 3.

logarithmic spiral A spiral traced out by a point rotating about a fixed point at constant speed while simultaneously moving away from the fixed point at a speed with acceleration proportional to its distance from the fixed point. It is given by the equation $r = e^{k\theta}$ in polar coordinates, with k

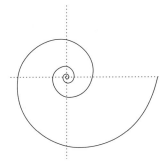

Logarithmic spiral

constant. This spiral is also called the spiral of Bernoulli, the exponential spiral, the equiangular spiral, and the spiral mirabilis.

logic Rules that govern the systematic deduction of valid conclusions from given assumptions.

longitude A semicircle on the surface of the Earth that is the intersection of the Earth with a half-plane whose edge is the north-south axis of the Earth.

loop A path whose starting point and ending point are the same.

loran A system of navigation for marine and air navigation that locates position based on the intersection of hyperbolas. Loran is short for "long-range navigation."

Lorentz transformation *See* HYPERBOLIC ROTATION.

Lorenz attractor *See* BUTTERFLY.

Lorenz butterfly *See* BUTTERFLY.

loxodrome A curve on the surface of a sphere that cuts the MERIDIANS at congruent angles as it spirals around the sphere.

loxodromic homography A product of four INVERSIONS or REFLECTIONS that cannot be expressed as the product of two inversions or reflections.

lozenge A RHOMBUS that is not a square.

lune The convex region between two intersecting circles in the plane or the region bounded by two great circles on a sphere.

lute of Pythagoras A progression of diminishing pentagons with inscribed pentagrams, a side of each pentagon being a side of the next pentagram.

m-equidissection The dissection of a polygon into m triangles with equal areas.

machinery *See* CONSTRUCTION.

Maclaurin series The POWER SERIES that most closely approximates a given function at its y-intercept.

magnitude Size. For example, the magnitude of a vector is its length.

main diagonal For a square matrix, the diagonal line of elements going from upper left to lower right.

major arc An arc of a circle that is larger than a semicircle.

major axis   In an ellipse, a segment through the foci connecting opposite points on the ellipse. It is the longer axis of symmetry of the ellipse.

mandala   The Sanskrit term for "circle," sometimes used to refer to a symmetric design with rotational symmetry.

Mandelbrot set   The set of complex numbers c such that the FILLED JULIA SET of the complex function $f(z) = z^2 + c$ is connected. Equivalently, the Mandelbrot set is set of numbers c such that c is in the PRISONER SET of the DYNAMICAL SYSTEM defined by $f(z) = z^2 + c$. It is a FRACTAL.

Manhattan metric   *See* TAXICAB METRIC.

manifold   A space in which every point is contained in a NEIGHBORHOOD that looks like a Euclidean space or half-space. Examples of manifolds are a TORUS and a sphere. The CANTOR SET is not a manifold.

map   (1) A partition of the sphere or other closed surface into faces by a GRAPH consisting of vertices and edges. (2) *See* FUNCTION.

map coloring   A coloring of a MAP in which regions sharing a boundary are colored differently.

mapping   *See* FUNCTION.

marked tiling   A tiling that has a design, called a marking or motif, on each tile. A symmetry of a marked tiling must also leave the markings or motifs apparently unchanged.

Mascheroni construction   A geometric construction that uses only a compass.

matrix   A rectangular array of numbers, which are called the elements or entries of the matrix. An $m \times n$ matrix has m rows and n columns. The plural is *matrices*.

maximal   As large as possible; having nothing bigger.

mean and extreme ratio   *See* GOLDEN RATIO.

mean curvature   The average of the two PRINCIPAL CURVATURES at a point on a surface.

mean proportional   For numbers a and b, the number x such that $a/x = x/b$.

means of a proportion   The numbers b and c in the proportion $a/b = c/d$.

measurement   A number giving the size or magnitude, such as length or area, of a geometric object or of some part of a geometric object.

mechanism   A LINKAGE, especially one for which there is only one point that is free to move independently.

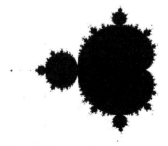

Mandelbrot set

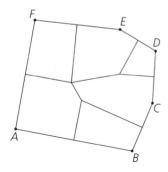

The medial axis of hexagon *ABCDEF*

medial area The area of a rectangle that is equal to an irrational number $\sqrt{a}$ where a is rational.

medial axis For a polygon the medial axis is the set of points in the interior of the polygon that have more than one closest point on the boundary of the polygon. For a set of points, the medial axis is the VORONOI DIAGRAM of the set.

medial rhombic triacontahedron *See* SMALL STELLATED TRIACONTAHEDRON.

medial segment A segment whose length is an irrational number $\sqrt[4]{a}$ where a is rational.

medial triangle The triangle whose vertices are the midpoints of the sides of a given triangle. It is similar to the original triangle.

median of a trapezoid A segment connecting the midpoints of the two nonparallel sides of a trapezoid. It is parallel to the two parallel sides of the trapezoid.

median of a triangle A segment connecting a vertex to the midpoint of the opposite side.

membership relation The RELATION that connects an element to a set in which it is contained, denoted by the symbol $\in$. Thus, $a \in S$ means that a is an element of the set S.

Menelaus point A point, other than a vertex, lying on the side or extended side of a triangle.

Menger sponge *See* SIERPINSKI SPONGE.

Mercator projection A PROJECTION of a sphere onto a cylinder such that the projections of LOXODROMES on the sphere become straight lines when the cylinder is unrolled.

meridian A curve on a surface of revolution that is the intersection of the surface with a plane containing the axis of revolution.

meridional disc A disc that is bounded by the meridian of a surface.

mesolabium An ancient mechanical device used to construct MEAN PROPORTIONALS.

metric A way of measuring the distance between two points. A metric has the following properties: The distance between a point and itself is 0; the distance from point A to point B is the same as the distance from point B to point A; and the distance from point A to point C is no bigger than the sum of the distance from point A to point B and the distance from point B to point C.

metric space A space with a METRIC giving the distance between any two points.

midcircle For two circles C_1 and C_2, a midcircle is a circle of INVERSION such that the inversive image of C_1 is C_2.

middle-C sequence A symmetric MUSICAL SEQUENCE found in a CARTWHEEL formed by PENROSE TILES.

midline *See* MEDIAN.

midpoint A point on a segment or arc equidistant from the two endpoints.

midradius The radius of a MIDSPHERE of a polyhedron.

midsphere A sphere that is tangent to all edges of a given polyhedron and whose center is at the center of the polyhedron.

mil One 6400th of a circle; a unit of angle measure used by the military.

Miller's solid *See* PSEUDORHOMBICUBOCTAHEDRON.

milligon *See* CENTESIMAL SECOND.

minimal As small as possible; having nothing smaller.

minimal curve A curve with the least arc length that satisfies a given set of conditions.

minimal surface A surface with the least area that satisfies a given set of conditions.

minimum dissection tiling A DISSECTION TILING of a polygon where the number of pieces of the polygon is the smallest possible.

Minkowski space The four-dimensional space used by physicists in special relativity.

Minkowski sum of two point sets A way of adding two sets in a coordinate space. The Minkowski sum of the two sets A and B of points in the coordinate plane is the set $\{(x, y) \mid x = a_1 + b_1, y = a_2 + b_2, \text{ for } (a_1, a_2) \text{ in } A, (b_1, b_2) \text{ in } B\}$.

minor arc An arc of a circle that is less than a semicircle.

minor axis In an ellipse, a segment perpendicular to a line through the foci connecting opposite points on the ellipse. It is the shorter axis of symmetry of the ellipse.

minute   One-sixtieth of a degree; a unit of angle measure.

Miquel circle   A Miquel circle passes through a vertex of a triangle and two arbitrary points, one on each of the sides adjacent to the vertex.

Miquel point   Choose three arbitrary points, one on each side of a triangle. A Miquel point is the intersection of the three MIQUEL CIRCLES that pass through a vertex of the triangle and the arbitrary points on the adjacent sides.

Miquel triangle   For a given point, any triangle that has the given point as a MIQUEL POINT.

MIRA   A piece of plastic that is translucent and reflective. It is used to study the reflections of geometric shapes.

mirror image   The image of an object under reflection.

mirror line   The line left fixed by a reflection in two-dimensional space.

mirror plane   A plane left fixed by reflection in three-dimensional space.

mirror symmetry   Reflective symmetry across a line or plane.

mixed triple product   *See* SCALAR TRIPLE PRODUCT.

Möbius band   A nonorientable one-sided surface obtained by gluing the ends of a rectangle to each other after a half-twist. It is also called a Möbius strip.

Möbius involution   A ROTARY HOMOGRAPHY that is the composition of two INVERSIONS with respect to two orthogonal circles.

Möbius map   *See* MÖBIUS TRANSFORMATION.

Möbius transformation   A complex function of the complex variable z of the form $f(z) = \dfrac{az + b}{cz + d}$ where $ad - bc \neq 0$. The image of a circle or line under a Möbius transformation is a circle or line.

Möbius triangle   A SPHERICAL TRIANGLE whose edges are the intersections of a sphere with the planes of symmetry of a UNIFORM POLYHEDRON inscribed in the sphere.

model   An interpretation of an AXIOMATIC SYSTEM in which all of the axioms are true. For example, the POINCARÉ DISC is a model for the axioms of HYPERBOLIC GEOMETRY.

modern compass   Dividers; a compass that can be tightened to maintain a fixed radius and loosened to change the radius.

modern elementary geometry   The study since about 1800 C.E. of circles and rectilinear figures based on Euclid's axioms.

Möbius band

modular surface　For a complex-valued function f of a complex variable $x + iy$, the surface consisting of all points of the form $(x, y, |f(x + iy)|)$.

modulus　The absolute value of a complex number.

Mohr-Mascheroni construction　*See* MASCHERONI CONSTRUCTION.

momentum vector　A VECTOR giving the magnitude and direction of the momentum of a moving mass.

Monge's form　An equation of the form $z = F(x, y)$ for a surface in three-dimensional space.

monkey saddle　A surface with a SADDLE POINT having three downward sloping regions (for the monkey's legs and tail) separated by three upward sloping regions.

monochord　A musical instrument with one string. It was used by Pythagoras to demonstrate the relationship between mathematics and music.

monogonal tiling　A tiling in which all VERTEX FIGURES are EQUIVALENT.

monohedral tiling　A tiling in which all tiles are congruent to each other.

monomorphic tile　A tile that has a unique MONOHEDRAL TILING of the plane.

Morse theory　The study of critical points of surfaces embedded in three-dimensional space.

mosaic　*See* TILING.

motif　A design that is used as a basis for a pattern or tiling of the plane.

motion　(1) An ISOMETRY of a shape or space onto itself, or in some contexts, a direct isometry of a shape or space onto itself. (2) A DEFORMATION of a FRAMEWORK or MECHANISM.

motion planning　Determining a path that a robot can take to avoid collisions. *See* FINE MOTION PLANNING and GROSS MOTION PLANNING.

moving trihedron　The FRAME consisting of the TANGENT VECTOR, the PRINCIPAL NORMAL VECTOR, and the BINORMAL VECTOR at a point on a curve. As the point on the curve changes, so does the moving trihedron.

multifoil　A design consisting of arcs of circles centered at the vertices of a regular polygon; the radius of the circles is half the side of the polygon and the arc is outside of the polygon with endpoints on the polygon.

multiple edges　Edges that connect the same two vertices of a GRAPH.

multiple point　A point of self-intersection of a curve.

multiplicity *See* DEGREE OF A ROOT.

music of the spheres The music that is believed to be generated by the harmonious relationships among the heavenly bodies as they travel. The Pythagoreans believed that the music of the spheres could be understood through mathematics.

musical sequence A sequence of intervals between AMMANN BARS found in a PENROSE TILING.

mutually visible points Points on a TERRAIN such that a segment joining any two of the points lies above the terrain.

Myreberg point The limit of the sequence of circles attached to the right of the main cardioid of the MANDELBROT SET in the complex plane. It is denoted by λ and is approximately equal to 1.40115.

myriagon A polygon with 10,000 sides.

mystic hexagram theorem Pascal's theorem that the intersections of opposite sides of a hexagon inscribed in a circle are collinear.

***n*-body problem** The problem of determining the trajectories of *n* planets moving in space under the force of gravitational attraction and Newton's Laws of Motion. No exact solution is known if *n* is greater than 2.

***n*-cage** A GRAPH having *n* vertices that has the smallest GIRTH of all graphs with *n* vertices.

***n*-connected** Describing a shape that will remain connected after $n-1$ cuts along nonintersecting SIMPLE CLOSED curves, but not after *n* such cuts. For example, a sphere is 1-connected and a TORUS is 2-connected.

***n*-dimensional** Describing a space with *n* DIMENSIONS, where *n* is a non-negative integer.

***n*-gon** A polygon with *n* vertices and *n* sides, for *n* greater than 2.

***n*-hex** A POLYHEX consisting of *n* hexagons.

***n*-iamond** A POLYIAMOND consisting of *n* triangles.

***n*-manifold** A MANIFOLD of dimension *n*. Every point is contained in a NEIGHBORHOOD that looks like *n*-dimensional Euclidean space.

***n*-omino** A polyomino that is the union of *n* squares.

***n*-polygon** *See n*-GON.

***n*-space** Euclidean space of *n* dimensions.

_n_-sphere A sphere in $(n + 1)$-dimensional space; it is an n-dimensional shape.

_n_-zonogon A polygon with $2n$ sides such that opposite sides are parallel. A convex n-zonogon is either a parallelogram or a hexagon.

_n_-zonohedron A ZONOHEDRON with n different zones.

nabla The symbol ∇ used to represent the GRADIENT FUNCTION.

Nagel point The point of concurrency of three lines, each of which connects a vertex of a triangle to a point halfway around the perimeter of the triangle from that vertex.

Napoleon triangle The equilateral triangle whose vertices are the centers of equilateral triangles constructed on the edges of a given triangle, either internally, giving the internal Napoleon triangle, or externally, giving the external Napoleon triangle.

nappe The entire sheet of a DOUBLE CONE that lies on one side of the vertex.

natural equation An equation for a curve in terms of its CURVATURE and TORSION, independent of a coordinate system. For example, a circle can be given by the natural equation $k = a$, where k is the curvature and a is a constant.

natural logarithm A logarithm for the base e.

nearest neighbor graph A GRAPH formed from a collection of points by connecting each point to the point or points closest to it.

necessary condition The CONCLUSION of an IMPLICATION. The necessary condition of the implication "$p \rightarrow q$" is the statement q.

negation A statement that asserts that another statement is false. For example, the negation of "Today is Monday" is the statement "Today is not Monday."

negative crossing A CROSSING of two directed curves in a KNOT DIAGRAM in which the lower curve goes from left to right as one travels along the upper curve.

neighborhood An OPEN SET. In some contexts, a neighborhood of a point is a set containing an open set that contains the point.

nephroid The EPICYCLOID with two CUSPS that is traced by a point on a circle rolling outside a circle with radius twice as large.

net A figure made up of many polygons that can be used to construct a polyhedron by folding along the shared edges and gluing appropriate pairs of edges.

Negative crossing

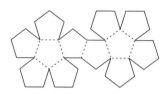

The net of a dodecahedron

net of curves A family of PARAMETRIC CURVES on a surface such that exactly two curves with different directions pass through each point. A net of curves can be used to define curvilinear coordinates on the surface.

net of rationality *See* HARMONIC NET. A harmonic net is called a net of rationality because the set of rationals is a harmonic net on the real number line.

network The GRAPH formed by the edges and vertices of a space-filling array of polyhedra.

neusis A construction technique using an idealized sliding ruler. Neusis constructions cannot be done with compass and straightedge, but can be used to solve the three classical problems.

neutral geometry *See* ABSOLUTE GEOMETRY.

Newton number *See* KISSING NUMBER.

nine-point circle The CIRCUMCIRCLE of the ORTHIC TRIANGLE of a given triangle. It passes through nine special points: the feet of the three altitudes, the midpoints of the three sides, and the midpoints of the segments joining the ORTHOCENTER to the three vertices.

node (1) A double point of a curve where there are two distinct tangents, one for each arc passing through the point. (2) A vertex of a graph.

non sequitur A statement that does not follow from a given line of reasoning.

non-Euclidean geometry ELLIPTIC OR HYPERBOLIC GEOMETRY.

nonagon A polygon with nine sides.

noncollinear Not lying on a common line; said of points.

noncommutative Not satisfying the COMMUTATIVE property. For example, subtraction and division are noncommutative.

noncommutative geometry The study of geometric properties that are associated with number systems having a noncommutative multiplication. Noncommutative geometry provides models of space that are used in the quantum theories of physics.

nonconvex Not CONVEX.

noncut point A point on a CONNECTED curve that is not a CUT POINT.

nondegenerate projective space A PROJECTIVE SPACE containing at least two lines.

nonempty Not empty.

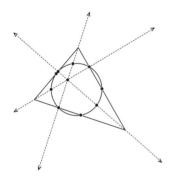

The nine-point circle of a triangle

nonintersecting line *See* ULTRAPARALLEL.

nonisometric transformation A TRANSFORMATION that is not an ISOMETRY.

nonorientable Not ORIENTABLE.

nonperiodic tiling A tiling of the plane that has no TRANSLATIONAL SYMMETRIES.

nonplanar graph A GRAPH that can be drawn in the plane only with self-intersections.

nonrectifiable curve A curve whose arc length cannot be determined. For example, the KOCH SNOWFLAKE is nonrectifiable because it is so jagged that its total arc length is infinite.

nonsecant A line that does not intersect a given curve.

nonsimple polygon A polygon in which two sides intersect at a point other than a vertex.

nonuniform Not uniform.

norm The length or magnitude of a VECTOR. The norm of the vector $\mathbf{v}$ is denoted $\|\mathbf{v}\|$.

normal A line perpendicular to the tangent of a curve or surface.

normal component The component of a vector perpendicular to a given vector.

normal indicatrix The image of a space curve on a unit sphere where the image of a point on the curve is the tip of the UNIT NORMAL vector to the curve at that point, displaced so its tail is at the center of the unit sphere.

normal mapping A map from a surface to three-dimensional space that maps a point on the surface to the tip of its NORMAL VECTOR.

normal plane For a point on a curve, the plane determined by the BINORMAL and the PRINCIPAL NORMAL.

null angle An angle equal to $0°$.

null vector *See* ZERO VECTOR.

nullspace *See* KERNEL.

number theory The investigation of properties of the positive integers, particularly with respect to multiplication.

oblate Flattened at the poles.

oblate spheroid The surface of revolution obtained when an ellipse is rotated about its MINOR AXIS.

oblique Not perpendicular.

oblique azimuthal projection An AZIMUTHAL PROJECTION that is not POLAR.

oblique coordinates Coordinates with respect to axes that are not perpendicular.

oblique cylinder A cylinder whose base is not orthogonal to its lateral surface.

oblique cylindrical projection A CYLINDRICAL PROJECTION that is neither REGULAR nor TRANSVERSE.

oblique Mercator projection The PROJECTION of the Earth's surface onto a cylinder that is tangent to the Earth near the center of a particular region.

oblong A shape that is longer than it is wide. For example, a rectangle that is not a square, an ellipse that is not a circle, and an ellipsoid that is not a sphere are all oblongs.

observation height The height of an observer above the ground.

obtuse angle An angle with measure greater than 90° but less than 180°.

obtuse golden triangle An isosceles triangle with base angles measuring 36°.

obverse of a knot The MIRROR IMAGE of a KNOT.

octacontagon A polygon with 80 sides.

octadecagon A polygon with 18 sides.

octagon A polygon with eight sides.

octagram The STAR POLYGON created from the octagon.

octahedral group The GROUP of DIRECT SYMMETRIES of an octahedron; it has 24 elements.

octahedron A polyhedron with eight congruent faces, each of which is an equilateral triangle. Sometimes, any polyhedron with eight faces.

octahedron elevatum FRA LUCA PACIOLI'S name for a STELLATED OCTAHEDRON.

octahemioctahedron A polyhedron with six quadrilateral faces and four hexagon faces.

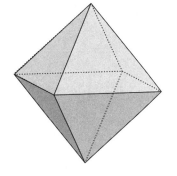

Octahedron

octant One-eighth of three-dimensional Cartesian space. The boundaries of the octants are the planes $x = 0$, $y = 0$, and $z = 0$.

octatetrahedron *See* OCTAHEMIOCTAHEDRON.

octet truss A space-filling polyhedron formed from two TETRAHEDRA and one OCTAHEDRON with congruent faces.

octomino A POLYOMINO made up of eight squares. There are 369 different octominos.

octonians A number system that is a generalization of the quaternions and contains 14 square roots of –1.

odd face A face of a polyhedron that has an odd number of sides.

odd function A function from the real numbers to the real numbers whose graph has POINT SYMMETRY about the origin. An odd function $y = f(x)$ has the property that $f(-x) = -f(-x)$. Examples of odd functions are $y = x^3$ and $y = \sin x$.

odd order horizon *See* BLOCKING SEGMENT.

odd permutation A permutation that is the product of an odd number of TRANSPOSITIONS.

odd vertex A vertex of a GRAPH incident to an odd number of edges.

omega triangle *See* ASYMPTOTIC TRIANGLE.

on Incident to.

one-dimensional form In PROJECTIVE GEOMETRY, a RANGE of points or a PENCIL of lines.

one-dimensional pattern *See* BAND ORNAMENT.

one-skeleton The figure formed by the edges and vertices of a polyhedron.

one-to-one correspondence A function from one set to another that is both one-to-one and onto.

one-to-one function A function that maps two distinct points in its domain to two distinct points in its range.

onto function A function that maps at least one point in its domain to each point in its range.

open ball The set of all points in the interior of a sphere.

open disc The set of all points in the interior of a circle.

open set A set whose complement is closed. Open sets have the following properties: The intersection of a finite number of open sets is open;

the union of an arbitrary number of open sets is open; and the empty set and the set containing every point in the space are open sets. *See* CLOSED SET.

opposite angles *See* VERTICAL ANGLES.

opposite of a statement *See* NEGATION.

opposite rays Two rays with a common endpoint that together form a line.

opposite sides In a COMPLETE QUADRANGLE, two lines that join disjoint pairs of points.

opposite symmetry A SYMMETRY of an oriented object that reverses the ORIENTATION.

opposite transformation A TRANSFORMATION that changes or reverses ORIENTATION.

opposite vertices Two vertices of a polygon that are separated by half of the sides of the polygon.

orbifold The IDENTIFICATION SPACE formed by identifying points of a shape that are mapped to one another by the elements of a GROUP of SYMMETRIES of the shape.

orbiform *See* CURVE OF CONSTANT WIDTH.

orbit The set of images of a point under a given set of TRANSFORMATIONS.

orbit analysis The process of determining the long-term behavior of the orbits of a dynamical system.

order Generally, the number of elements in a set.

order of a graph The number of vertices in a GRAPH.

order of a group The number of elements in the GROUP.

order of a matrix The number of rows or columns in a square matrix.

order of a projective geometry One less than the number of points on any line in the geometry.

order of a root For a root r of a given polynomial, the highest power of $(x - r)$ that divides the polynomial.

order of a symmetry The smallest number of times a symmetry can be composed with itself and give the identity.

order of an affine space The number of points in any line in the space.

ordered geometry The study of INTERMEDIACY and its properties.

ordering A relationship on the elements of a set such that for any two distinct elements a and b either $a > b$ or $b > a$.

ordinary point *See* REGULAR POINT.

ordinate The second, or y-, coordinate of a point.

orientable Describing a curve or surface that has a continuous choice of TANGENT VECTOR or NORMAL VECTOR, respectively, at each point. The sphere is orientable, while the MÖBIUS BAND and the KLEIN BOTTLE are not orientable.

orientation A direction. For a curve or surface, an orientation is given by a TANGENT VECTOR or NORMAL VECTOR, respectively, at a point. An orientation of a vector space is an ordering of the basis vectors.

oriented knot A KNOT together with an ORIENTATION.

oriented link A LINK together with an ORIENTATION for each of its COMPONENTS.

oriented simplex A SIMPLEX with an ordering of its vertices.

origami The Japanese art of folding paper to create different shapes. Origami can be used to construct many geometric shapes, including the cube.

origin In a coordinate system, the point all of whose coordinates are 0.

ornamental group The SYMMETRY GROUP of a plane design or pattern. The ornamental groups are the rosette groups, the frieze groups, and the wallpaper groups.

orthant One of the regions determined by the coordinate axes in a higher-dimensional space. In n-dimension space, there are 2^n orthants.

orthic axis The TRILINEAR POLAR of the ORTHOCENTER of a triangle.

orthic triangle A triangle whose vertices are the feet of the altitudes of a given triangle.

orthocenter The intersection of the three altitudes of a triangle.

orthocentric quadrangle A COMPLETE QUADRANGLE whose three pairs of opposite sides are perpendicular. The vertices of an orthocentric quadrangle form an ORTHOCENTRIC SET.

orthocentric set A set of four points such that each point is the ORTHOCENTER of the triangle formed by the other three points.

orthocentroidal circle A circle whose diameter has endpoints at the CENTROID and the OTHOCENTER of a triangle.

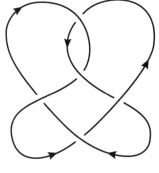

Oriented knot

orthodrome *See* GREAT CIRCLE.

orthogonal *See* PERPENDICULAR.

orthogonal curves Two curves with the property that, at each point of intersection of the two curves, the tangent lines are perpendicular.

orthogonal graph A PLANAR GRAPH that can be drawn so that all edges consist only of horizontal and vertical segments.

orthogonal matrix A square matrix whose rows (or columns) are mutually orthogonal when considered as VECTORS. An orthogonal matrix corresponds to an ISOMETRIC LINEAR TRANSFORMATION.

orthogonal net Two families of curves with the property that each curve of one family is orthogonal to every curve of the other family.

orthogonal projection A PROJECTION of a set onto a given plane that maps each point in the domain to the foot of the perpendicular dropped from the point to the plane.

orthogonal trajectory A curve that is orthogonal to each curve in a given family of curves.

orthogonal transformation A LINEAR TRANSFORMATION that preserves DOT PRODUCTS, angle measure, and NORMS. It can be represented by an ORTHOGONAL MATRIX.

orthographic projection *See* ORTHOGONAL PROJECTION.

orthopole For each vertex of a given triangle, drop a perpendicular to a given line and drop a perpendicular from the foot of that perpendicular to the side opposite the vertex. The point of concurrency of the three lines constructed in this way is the orthopole of the triangle.

orthoptic curve An ISOPTIC CURVE whose points are the intersections of tangents that meet at right angles.

orthoscheme *See* QUADRIRECTANGULAR.

orthotope A PARALLELOTOPE whose edges meet at right angles.

osculating circle *See* CIRCLE OF CURVATURE.

osculating plane The plane through the TANGENT LINE and the PRINCIPAL NORMAL VECTOR at a point on a curve. It contains the OSCULATING CIRCLE at that point.

osculating sphere The sphere that fits a curve at a given point better than any other sphere. Its center is the center of curvature and its radius is the radius of curvature.

osculation, point of *See* TACNODE.

osculinflection A point of INFLECTION that is also a point of OSCULATION.

outer product *See* CROSS PRODUCT.

oval A SIMPLE CLOSED PLANAR CURVE that is SMOOTH and CONVEX and at every point the CURVATURE VECTOR points inward.

ovaloid A CONVEX CLOSED SURFACE that has continuous nonvanishing PRINCIPAL CURVATURES and positive TOTAL CURVATURE.

overcrossing The arc of a KNOT above a CROSSING in the KNOT DIAGRAM.

overdetermine To place so many conditions on the construction of an object that it is impossible to construct it.

overpass *See* OVERCROSSING.

ovoid The surface generated by rotating an oval about its axis of symmetry.

{*p, q*} The SCHLÄFLI SYMBOL for a polyhedron or tiling whose faces or tiles are regular p-GONS, with q of them meeting at each vertex.

(*p, q*) *See* {p, q}.

{*p, q, r*} The SCHLÄFLI SYMBOL for a regular polytope whose cells are regular {p, q} polyhedra and whose VERTEX FIGURES are regular {q, r} polyhedra.

(*p, q*)-torus knot For relatively prime integers p and q, a (p, q)-torus knot is a KNOT that can be drawn on the surface of a TORUS so that it passes through the hole in the torus p times as it winds around the torus q times, never crossing itself.

(*p*λ) The symbol used to represent a CONFIGURATION with p points and p lines, with each point on λ lines and each line containing λ points.

(*p*λ/*l*π) The symbol used to represent a CONFIGURATION with p points and l lines, with each point on λ lines and each line containing π points. The condition $p\lambda = l\pi$ must hold.

$\{\frac{p}{q}\}$ The symbol used to denote the STAR POLYGON obtained by connecting every qth vertex of a regular p-gon.

p-gon A polygon with p vertices and p sides.

packing A collection of subsets of a region that are disjoint or intersect only along their boundaries. The sets in a packing may or may not cover the whole region.

pangeometry The term used by LOBACHEVSKY for HYPERBOLIC GEOMETRY.

pantograph A mechanical device used to draw a reduction or enlargement of a figure.

Pappian projective space A PROJECTIVE SPACE in which the theorem of PAPPUS is valid.

Pappus, finite geometry of A finite geometry with exactly nine points and nine lines. Each point lies on exactly three lines.

Pappus configuration A SELF-DUAL CONFIGURATION with nine lines and nine points such that each line contains three points.

par-hexagon A hexagon whose opposite sides are parallel. Any par-hexagon will tile the plane.

parabola The locus of all points equidistant from a line, called the directrix of the parabola, and a point, called the focus of the parabola. *See also* CONIC SECTION.

parabolic curve A curve on a surface such that the GAUSSIAN CURVATURE at each point is equal to 0.

parabolic cylinder The surface swept out by a parabola translated along a line perpendicular to the plane of the parabola.

parabolic geometry Euclidean geometry, a term originated by FELIX KLEIN to distinguish Euclidean geometry from the two non-Euclidean geometries, ELLIPTIC GEOMETRY and HYPERBOLIC GEOMETRY.

parabolic homography The product of two INVERSIONS with respect to tangent circles.

parabolic point A point on a surface where the intersection of the tangent plane with the surface is a curve. The GAUSSIAN CURVATURE at a parabolic point is 0. For example, any point on a cylinder is a parabolic point.

parabolic projectivity *See* PROJECTIVITY.

parabolic rotation An EQUIAFFINITY of the form $f(x, y) = (x + 1, 2x + y + 1)$.

parabolic umbilic A PARABOLIC POINT that is an UMBILIC.

paraboloid The surface formed when a parabola is rotated about its axis.

paradromic rings Rings of equal width that are produced by cutting a strip lengthwise that has been given any number of half twists and then attached at its ends.

parallel (1) In general, *parallel* describes objects that do not intersect. Two or more lines in the same plane are parallel if they do not intersect; a

Parabola

line and a plane in three-dimensional space are parallel if they do not intersect; and two or more planes in three-dimensional space are parallel if they do not intersect. Figures in parallel planes are parallel. *See also* SKEW LINES. (2) A curve on a surface of revolution that is the intersection of the surface with a plane orthogonal to the axis of revolution. This curve is also called a parallel circle.

parallel angles CORRESPONDING ANGLES for a TRANSVERSAL that crosses parallel lines.

parallel class The set of all lines parallel to a given line.

parallel curve A curve each of whose points is at a constant distance, measured along a NORMAL, from another curve.

parallel displacement *See* PARALLEL TRANSPORT.

parallel line In HYPERBOLIC GEOMETRY, one of the two lines through a given point that are closest to a given line and do not intersect the given line.

parallel postulate Euclid's postulate that ensures the existence of parallel lines.

parallel projection The PROJECTION of points in three-dimensional space onto an image plane along a fixed direction.

parallel transport Motion of a VECTOR along a curve so that its tail is on the curve and the angle between the vector and the tangent to the curve is constant.

parallelepiped A polyhedron with three pairs of parallel faces; each face is a parallelogram.

parallelepiped product *See* TRIPLE SCALAR PRODUCT.

parallelepipedon *See* PARALLELEPIPED.

parallelism The quality or property of being parallel.

parallelogon A polygon with opposite sides parallel to each other. A parallelogon is either a parallelogram or a hexagon.

parallelogram A quadrilateral in which both pairs of opposites sides are parallel.

Parallelogram Law The rule that says vector addition is commutative. The name comes from the vector parallelogram that shows $\mathbf{v} + \mathbf{w} = \mathbf{v} + \mathbf{w}$ for vectors $\mathbf{v}$ and $\mathbf{w}$.

parallelohedron A ZONOHEDRON that fills space. There are five parallelohedra: the cube, hexagonal prism, rhombic dodecahedron, elongated dodecahedron, and truncated octahedron.

parallelotope A polytope whose faces are PARALLELEPIPEDS.

parameter An independent variable, especially the variable t when used to represent time.

parameter curve A curve described by PARAMETRIC EQUATIONS.

parameterization with respect to arc length Describing a curve as a parameter curve so that the change in the value of the parameter t between any two points on the curve is equal to the length of the curve between those two points.

parametric equation An equation giving a dependent variable in terms of one or more PARAMETERS. For example, the pair of parametric equations $x = \cos t$ and $y = \sin t$ describes a circle in the Cartesian plane.

parasphere *See* HOROSPHERE.

parquetting *See* TILING.

partial derivative A measure of the rate of change of a function of several variables with respect to one of the variables.

partial differential equation An equation that uses PARTIAL DERIVATIVES to describe how changing quantities are related to one another.

partial sum *See* SERIES.

Pascal line A line containing the points of intersections of the pairs of opposite sides of a hexagon inscribed in a circle or other CONIC SECTION. Six points lying on a conic section determine 60 different hexagons and these determine 60 Pascal lines.

Pasch's axiom An axiom added to Euclidean geometry by MORITZ PASCH in 1882. It says that, in the plane, a line intersecting one side of a triangle and but none of the three vertices must intersect one of the other two sides. In a more general form, it states that a line divides the plane containing it into two disjoint half-planes and that a plane divides the space containing it into two disjoint half-spaces.

pass A SADDLE POINT of an ALTITUDE FUNCTION on a sphere.

patch A region of a surface bounded by a simple closed curve.

path A curve, especially one parameterized by the PARAMETER t.

pathwise connected *See* ARCWISE CONNECTED.

pattern recognition The problem in computer vision of determining the nature of an object from specific measurements of the object.

paving *See* TILING.

PC-point *See* PEDAL-CEVIAN POINT.

peak A point where there is a local maximum value of an ALTITUDE function on a sphere.

Peano curve A fractal curve that passes through every point in the interior of a square.

Peaucellier's linkage A mechanical device consisting of six rods used to draw a straight line. It is also called Peaucellier's inversor or Peaucellier's cell.

pedal-cevian point A point whose PEDAL TRIANGLE for a given triangle is a CEVIAN TRIANGLE of that triangle.

pedal curve A curve consisting of the feet of the perpendiculars dropped from a given point to each of the lines tangent to a given curve.

pedal line *See* SIMSON LINE.

pedal-point construction Finding a curve that will obtain a given curve as a PEDAL CURVE for a given point.

pedal triangle *See* ORTHIC TRIANGLE.

pencil of coaxial circles A family of circles with the same LINE OF CENTERS and the same RADICAL AXIS. If there are two points common to every circle, the pencil is elliptic or intersecting; if the circles are disjoint, the pencil is hyperbolic or nonintersecting; and if there is one point common to every circle in the pencil, the pencil is parabolic or tangent.

pencil of lines A set of concurrent lines or a set of parallel lines.

pencil of planes A set of planes that intersect in a line or a set of parallel planes.

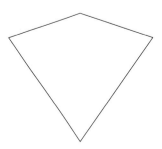

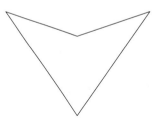

Penrose kite and dart

pencil of ultraparallels In HYPERBOLIC GEOMETRY, all lines perpendicular to a given line.

Penrose kites and darts *See* PENROSE TILES.

Penrose rhombs Two MARKED rhombus PROTOTILES that tile the plane only aperiodically.

Penrose tiles Two MARKED PROTOTILES, one shaped like a kite and one shaped like a dart, that tile the plane only aperiodically.

Penrose tiling A tiling by PENROSE TILES.

Penrose tiling

Penrose's wheelbarrow *See* WHEELBARROW.

pentacaidecahedron A polyhedron with 15 faces.

pentacle *See* PENTAGRAM.

pentacontagon A polygon with 50 sides.

pentacontahedron A polyhedron with 50 faces.

pentadecagon A polygon with 15 sides.

pentagon A polygon with five sides.

pentagonal dipyramid A CONVEX polyhedron with 10 triangular faces and seven vertices. It is a DELTAHEDRON and is the DUAL of the pentagonal prism.

pentagonal hexecontahedron A polyhedron with 60 pentagonal faces and 92 vertices. It is the DUAL of the snub dodecahedron.

pentagonal icositetrahedron A polygon with 24 pentagonal faces and 38 vertices. It is the dual of the SNUB CUBE.

pentagram The star-shaped figure consisting of the five diagonals of a regular pentagon. It is the STAR POLYGON obtained from a pentagon.

pentagrid A superposition of five symmetric GRIDS used in the study of PENROSE TILINGS.

pentahedron A polyhedron with five faces.

pentakaidecahedron A polyhedron with 15 faces.

pentakis dodecahedron A polyhedron with 60 triangular faces and 32 vertices. It is the dual of the truncated icosahedron.

pentatope *See* 5-CELL.

pentiamond A POLYIAMOND made up of five equilateral triangles.

pentoil knot The (2, 5) TORUS KNOT; it has crossing number 5.

pentomino A POLYOMINO made up of five squares. For example, the GREEK CROSS is a pentomino.

perceptron In computer learning, a model used to classify data points into two sets.

perfect coloring A coloring of a pattern so that all of the symmetries of the colored pattern are COLOR SYMMETRIES.

perfect cuboid An EULER BRICK with SPACE DIAGONAL that has integer length; no perfect cuboids are known to exist.

perfectly colored Describing a colored pattern or design all of whose symmetries are COLOR SYMMETRIES.

perimeter The sum of the lengths of the sides of a polygon.

period (1) For a PERIODIC FUNCTION f, the period is the smallest number p such that $f(x) = f(x + p)$. (2) For a PERIODIC POINT of a function, the period is the smallest number of iterations of the function that will bring the point back to its original position. (3) For any function, the period is number of times it must be composed with itself to give the identity; for example, $f(x) = -x$ has period 2 and rotation by $120°$ about a given point has period 3. (4) For an element of a GROUP, the period is the smallest positive number so that the element raised to that power is the identity. If there is no finite number that is the period of a function or an element, it is said to have infinite period.

period doubling The phenomenon in some discrete DYNAMICAL SYSTEMS that the number of points in a stable cycle doubles when the values of the parameters of the system are changed appropriately.

period parallelogram For a LATTICE, a parallelogram whose vertices are points of the lattice and which does not contain any LATTICE POINTS.

periodic function A function whose values keeps repeating. A periodic function f has the property that $f(x) = f(x + p)$ for some p and all x; the smallest such p for which this equation is valid is called the period of the function.

periodic group *See* WALLPAPER GROUP.

periodic point A point that returns to its original position after a finite number of iterations of a function.

periodic tiling A tiling which has TRANSLATIONAL SYMMETRY in two different directions.

peripheral angle *See* INSCRIBED ANGLE.

periphery The boundary of a planar region.

Perko pair A pair of equivalent KNOTS with nine CROSSINGS. For many years, they were thought to be inequivalent.

permutation A one-to-one correspondence of a set with itself; a rearrangement of a set.

perpendicular Meeting at right angles.

perpendicular bisector The perpendicular bisector of a segment is a perpendicular line passing through the midpoint of the segment.

perspective (1) The science of drawing a picture on a two-dimensional surface so that it appears three-dimensional to the viewer. (2) Two points on two different lines are perspective if there is a PERSPECTIVITY from the line containing one of the points to the line containing the other point that maps one point to the other.

perspective collineation A COLLINEATION of a PROJECTIVE SPACE which leaves fixed all the lines through a point, called the center of the perspective collineation, and all the points on a line, called the axis of the perspective collineation.

perspective from a line Two polygons are perspective from a line *l* if corresponding sides of the polygons meet at points on the line *l*.

perspective from a point Two polygons are perspective from a point *P* if lines joining corresponding vertices are concurrent at the point *P*.

perspective projection The PROJECTION of points in three-dimensional space onto a two-dimensional image plane from a fixed point, called the center of projection. The image of a point is the intersection of the image plane with the line connecting the point to the center of projection.

perspectivity A one-to-one correspondence between the points on two lines determined by a given point, not on either line, called the center of perspectivity or point of perspectivity. A point on one line, its corresponding point on the other line, and the center of perspectivity are collinear. A perspectivity between two planes in three-dimensional space is defined similarly.

Petrie polygon A SKEW POLYGON associated with a regular polyhedron such that every two, but no three, consecutive sides, are edges of the same face of the polyhedron.

PG(n, q) A FINITE PROJECTIVE GEOMETRY with dimension *n* and order *q*.

phase *See* ARGUMENT OF A COMPLEX NUMBER.

phase portrait A diagram that shows all possible states and evolutionary paths of a DYNAMICAL SYSTEM, including the EQUILIBRIUM POINTS of the system, the BASINS and their ATTRACTORS, and the lines separating the basins.

pi (π) The ratio of the circumference of a circle to the diameter of a circle. It is approximately equal to 3.14159.

picture plane *See* IMAGE PLANE.

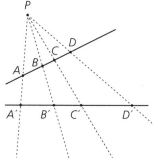

Perspectivity with center of perspectivity *P*

pit A point where there is a local minimum value of an ALTITUDE function on a sphere.

pitch (1) A measure of how much a ruled surface bends or curves in space. It is the measure of the angle between two GENERATORS divided by the distance between them as the generators get very close to each other. (2) The change in height of a point on a helix corresponding to a complete counterclockwise revolution about the axis of the helix.

pixel A small patch of a computer screen that can be given any one of a number of different colors.

PL-topology Piece-wise linear topology; *see* COMBINATORIAL TOPOLOGY.

planar Lying in a plane.

planar graph A graph that can be drawn in a plane.

planar net of a polyhedron *See* SCHLEGEL DIAGRAM.

plane A flat two-dimensional space. In Euclidean geometry, the plane is infinite and has no boundary.

plane at infinity *See* IDEAL PLANE.

plane crystallographic group *See* WALLPAPER GROUP.

plane curve A curve that lies in a plane.

plane dual The dual of a statement in two-dimensional PROJECTIVE GEOMETRY obtained by replacing each occurrence of the word *point* by *line* and each occurrence of the word *line* by the word *point*.

plane geometry The EUCLIDEAN GEOMETRY of the plane.

plane of symmetry *See* MIRROR PLANE.

plane polygon A polygon lying in a plane. Its vertices are coplanar.

plane surveying The surveying of small areas when it is not necessary to take into account the curvature of the Earth's surface.

planimeter A mechanical device used to measure the area of a planar region.

plate and hinge framework A FRAMEWORK consisting of flat rigid polygonal faces joined by flexible hinges.

Plateau's problem The problem of finding a surface of minimal area that has a given nonplanar simple closed curve as its boundary. The solution to Plateau's problem corresponds to a soap film across a wire bent in the shape of the given curve.

Platonic solid The five regular polyhedra: tetrahedron, octagon, icosahedron, cube, and dodecagon; so called because PLATO describes them in the *Timaeus*.

plot To construct the graph of a point or equation using a coordinate system.

Plücker coordinates HOMOGENEOUS COORDINATES for a line in a three-dimensional projective space.

Plücker line A line passing through four STEINER POINTS. Any six points lying on a conic section determine 15 Plücker lines.

Poincaré conjecture The conjecture that a SIMPLY CONNECTED CLOSED three-dimensional MANIFOLD is topologically equivalent to a three-dimensional sphere. This conjecture has not been proved, although its generalization to higher-dimensional manifolds has been proved.

Poincaré disc A model of HYPERBOLIC GEOMETRY consisting of the interior of a circle; the lines of the geometry are the diameters of the circle and arcs of circles that are orthogonal to the circle.

Poincaré group *See* FUNDAMENTAL GROUP.

Poincaré half-plane A model of HYPERBOLIC GEOMETRY that uses a metric different from the usual Euclidean metric on the half-plane where $y > 0$.

point The most fundamental unit in a geometry, used to mark a location in space. A point cannot be broken down into parts and has no length or width. *Point* is usually an undefined term in a geometry. Following EUCLID, points are usually represented by a capital letter.

point at infinity *See* IDEAL POINT.

point conic A point conic is the reformulation of the concept of a CONIC SECTION into the language of PROJECTIVE GEOMETRY. A point conic is the set of points of intersection of corresponding lines in two PENCILS that are related by a PROJECTIVITY but not a PERSPECTIVITY.

point group *See* CRYSTALLOGRAPHIC POINT GROUP.

point lattice *See* LATTICE.

point of application The initial point of a vector.

point of contact *See* POINT OF TANGENCY.

point of osculation A DOUBLE POINT on a curve where the two arcs of the curve have the same tangent line.

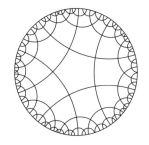

Poincaré disc

point of perspectivity *See* PERSPECTIVITY.

point of projectivity *See* PROJECTIVITY.

point of symmetry The center of a POINT SYMMETRY.

point of tangency The point where a tangent line or plane is tangent to a curve or surface.

point reflection *See* POINT SYMMETRY.

point row All the points on a line.

point set A set whose elements are points.

point source In a DYNAMICAL SYSTEM, a SOURCE that is a point.

point symmetry An ISOMETRY that, for a fixed point O called the center of the point symmetry, maps a point P to a point P' such that O is the midpoint of segment PP'. In the plane, a point symmetry is a half-turn. In three-dimensional space, a point symmetry is an opposite symmetry.

point-group symmetry Having only ROTATIONAL or REFLECTIONAL SYMMETRY.

Point-Line Incidence Axiom *See* LINE AXIOM.

point-point visibility function *See* VISIBILITY FUNCTION.

point-region visibility function *See* VISIBILITY FUNCTION.

point-set topology The study of TOPOLOGICAL SPACES using concepts and tools from set theory.

pointer A point with an associated direction, usually shown by a short arrow. For example, a 3-center can have three pointers attached to show the degree of rotational symmetry at that point.

pointwise inversion *See* MÖBIUS INVERSION.

polar (1) in general, a polar is a line with a special relationship to a specific point. The polar of a point with respect to a CIRCLE OF INVERSION is a line through the INVERSE of the given point that is perpendicular to the line connecting the point to the center of the given circle. A line is a polar of a given point if it is the LINE OF PERSPECTIVITY for two triangles having the given point as point of perspectivity. In ELLIPTIC GEOMETRY, the polar of a point is the line perpendicular to all lines passing through the point. (2) The line that is the image of a point under a POLARITY.

polar angle In a polar, cylindrical, or spherical coordinate system, the polar angle is the angle between the polar axis and a ray to a point.

polar axis (1) For a point on a curve, the polar axis is a line passing through the CENTER OF CURVATURE and perpendicular to the OSCULATING PLANE. (2) In a polar, cylindrical, or spherical coordinate system, the ray from which angles are measured.

polar azimuthal projection An AZIMUTHAL PROJECTION of a sphere to a plane tangent to the sphere at a pole.

polar circle For a SELF-POLAR TRIANGLE, the circle with respect to which the triangle is transformed to itself under RECIPROCATION.

polar coordinates Coordinates (r, θ) for a point in the plane where r is the distance of the point from the origin and θ gives the directed angle from the horizontal axis to the ray connecting the origin to the point.

polar distance *See* CODECLINATION.

polar form For a complex number $a + bi$, its expression in the form $r (\cos \theta + i \sin \theta)$.

polar line *See* POLAR AXIS.

polar plane The polar plane of a given point with respect to a sphere is the plane that passes through the INVERSE of the point with respect to the sphere and that is perpendicular to the line connecting the center of the sphere to the point.

polar triangle A triangle in which each vertex is the POLE of the opposite side.

polar triangles *See* CONJUGATE TRIANGLES.

polarity A PROJECTIVE CORRELATION that is its own inverse. A hyperbolic polarity has a point that lies on its image under the polarity and an elliptic polarity has no such point.

pole (1) In general, a pole is a point with a special relationship to a line, curve, or surface. The pole of a line with respect to a CIRCLE OF INVERSION is the INVERSE of the point of intersection of the given line with a perpendicular dropped from the center of inversion. A point is a pole of a given line if it is the POINT OF PERSPECTIVITY for two triangles having the given line as LINE OF PERSPECTIVITY. In ELLIPTIC GEOMETRY, the pole of a line is the intersection of all perpendiculars to the line. (2) The image of a line under a POLARITY. (3) The origin in a polar coordinate system. It is the endpoint of the polar axis. (4) One of the two endpoints of a diameter of a sphere.

polycube A solid made by joining congruent cubes along their faces.

polygon A planar region bounded by segments. The segments bounding the polygon are the sides of the polygon and their endpoints are the vertices of the polygon.

polygonal curve A curve consisting of segments.

polygonal knot A KNOT that is made up of a finite number of segments.

polygonal region *See* POLYGON.

polyhedral angle A three-dimensional figure formed by a vertex of a polyhedron and its adjacent faces.

polyhedral surface A surface that is made up of a finite number of triangles with at most two triangles sharing a given edge and such that the boundary of the surface is a disjoint union of edges.

polyhedroid A higher-dimensional region bounded by HYPERPLANES.

polyhedron A solid bounded by polygons. The polygons bounding the polyhedron are the faces of the polyhedron, their sides are the edges of the polyhedron, and their vertices are the vertices of the polyhedron.

polyhex A polygon that is the union of regular hexagons joined along their edges.

polyiamond A polygon that is the union of equilateral triangles joined along their edges.

polynomial An algebraic expression that is the sum of products of numbers and variables. For example, $3x^2 + 4xy^3$, $3x$, and 5 are all polynomials.

polyomino A polygon that is the union of squares joined along their edges.

polytope A convex polyhedron. In some contexts, a polytope is a higher-dimensional analogue of a polyhedron.

Poncelet transverse For a circle that lies inside another circle, the Poncelet transverse is a sequence of chords of the larger circle that are tangent to the smaller circle such that each chord shares an endpoint with the next chord in the sequence. An n-sided Poncelet transverse consists of n chords.

pons asinorum The Euclidean theorem stating that the two base angles of an isosceles triangle are equal.

porism A THEOREM or proposition that can be derived from the proof of another theorem.

position vector A VECTOR whose terminal point corresponds to the location of a specific point under consideration and whose initial point is the origin.

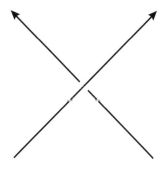

Positive crossing

positive crossing A crossing of two directed arcs in a KNOT DIAGRAM in which the lower curve passes from right to left as one travels along the upper curve.

positive grip A grip by a robot hand on an object that ignores friction between the robot's fingers and the object.

positive hull The smallest POSITIVE SET containing a given set of points.

positive linear combination A sum of SCALAR MULTIPLES of one or more VECTORS where the scalars are non-negative.

positive sense The direction along a PARAMETRIC CURVE in which the parameter increases.

positive set A set containing the POSITIVE LINEAR COMBINATIONS of its points.

postulate An AXIOM, especially one that applies to the specific branch of mathematics being considered.

Postulate of Line Measure *See* RULER AXIOM.

potential The SCALAR FUNCTION whose GRADIENT FIELD is a given CONSERVATIVE FIELD.

power The power of a point with respect to a circle is the square of the distance from the point to the center of the circle minus the square of the radius of the circle.

power of inversion The square of the radius of a CIRCLE OF INVERSION.

power series An infinite sum of non-negative powers of a variable. Thus, $1 + x + x^2 + x^3 + \ldots$ is a power series.

prefractal A geometric shape that is not itself a fractal but is the result of a finite number of steps of an iterative process that leads to the creation of a fractal.

preimage The preimage of a given point in the image of a function is a point in the domain of the function that maps to the given point.

pretzel A sphere with two or more HANDLES attached.

prime edges The prime edges of a k-dimensional PARALLELOTOPE in a k-dimensional AFFINE SPACE are k distinct edges of the parallelotope sharing a common endpoint.

prime knot A KNOT that is not the CONNECTED sum of two nontrivial knots.

prime number A positive integer other than 1 that cannot be factored except as itself times 1. For example, 2, 3, 5, 7, and 11 are prime.

primitive cell A LATTICE UNIT that does not contain any LATTICE POINTS other than its vertices.

primitive concept *See* UNDEFINED TERM.

primitive hypercube lattice The LATTICE in n-dimensional space containing all points whose rectangular coordinates are integers.

primitive lattice The simplest lattice in a CRYSTAL SYSTEM. There are seven primitive lattices, one for each crystal system.

primitive parallelogram A PRIMITIVE CELL that is a parallelogram.

primitive pattern A pattern whose only SYMMETRY is the identity.

primitive Pythagorean triangle A PYTHAGOREAN TRIANGLE in which the lengths of the three sides are relatively prime.

primitive Pythagorean triple A PYTHAGOREAN TRIPLE in which the three integers are relatively prime.

primitive term *See* UNDEFINED TERM.

principal bundle A FIBER BUNDLE whose FIBER is a GROUP.

principal curvature of a surface At a point on a surface, the maximum or minimum value of CURVATURE over all the NORMAL SECTIONS at that point.

principal diagonal *See* MAIN DIAGONAL.

principal direction At a point on a surface, a principal direction is a direction in which the RADIUS OF CURVATURE is either a maximum or a minimum. Except where the radius of curvature is constant, a point will have two principal directions.

principal normal The line determined by a PRINCIPAL NORMAL VECTOR.

principal normal vector A UNIT VECTOR NORMAL to a curve or surface that points toward the center of the CIRCLE OF CURVATURE or the SPHERE OF CURVATURE.

principal value (1) The nth root of a complex number with the smallest argument. (2) The value in the interval $(-180°, 180°]$ for the measure of the angle between the positive real axis and a ray from the origin to the graph of a complex number.

prism A polyhedron with two congruent parallel bases joined by lateral faces that are rectangles or parallelograms.

prismatoid A polyhedron with two parallel bases that are joined by lateral faces that are triangles, trapezoids, or parallelograms.

prismoid A PRISMATOID that has bases with the same number of sides joined by lateral faces that are trapezoids or parallelograms. In some contexts, a prismoid is an ANTIPRISM.

prisoner set All points whose images stay in a bounded region under the iterations of a DYNAMICAL SYSTEM.

Procrustean stretch *See* HYPERBOLIC ROTATION.

product of functions *See* COMPOSITION.

profile The intersection of a surface of revolution with a half-plane whose edge is the axis of revolution.

project To map using a PROJECTION.

projection A function from one surface to another surface. In linear algebra, a projection is a linear transformation f such that $f^2 = f$.

projection plane A plane that is the range space of a PROJECTION.

projective closure The smallest PROJECTIVE SPACE containing a given AFFINE SPACE.

projective collineation A COLLINEATION that is a PROJECTIVITY when restricted to any line in its domain.

projective correlation A CORRELATION that takes collinear points to concurrent lines and takes concurrent lines to collinear points.

projective geometry A geometry in which any two points determine a unique line and any two lines intersect at unique point. In projective geometry, incidence relations are most important. Projective geometry provides the theoretical basis for the science of perspective.

projective mapping *See* PROJECTION.

projective plane A nondegenerate two-dimensional PROJECTIVE SPACE.

projective space A space consisting of points, lines, and planes such that any two distinct points determine a line, a line intersecting two sides of a triangle will also intersect the third, and any line contains at least three points. In a projective space, the most important relationship is INCIDENCE.

projectivity The composition of two or more PERSPECTIVITIES. A projectivity is elliptic, parabolic, or hyperbolic if the number of fixed points is 0, 1, or 2, respectively.

projector line The line connecting a WORLD POINT to its image point under a given PROJECTION.

projector plane The plane consisting of all PROJECTOR LINES belonging to the points on a given line under parallel projection.

prolate Elongated along the polar axes.

prolate cycloid *See* CYCLOID.

prolate ellipsoid The surface of revolution obtained when an ellipse is rotated about its MAJOR AXIS.

proof A logical deduction of a theorem from axioms and previously proved results.

proof by contradiction A proof of a statement in which a contradiction is derived from the negation of the statement. If the negation of a statement leads to a contradiction, then the negation must be false and therefore the statement itself must be true.

proper motion *See* DIRECT ISOMETRY.

proper tiling A tiling by polygons such that the intersection of any two tiles is contained in an edge of each.

proportion An equality of ratios, $a/b = c/d$. A proportion can also be written $a:b = c:d$ or $a:b :: c:d$.

proportion by addition The proportion $\dfrac{a + b}{b} = \dfrac{c + d}{d}$ obtained from the proportion $a/b = c/d$.

proportion by addition and subtraction The proportion $\dfrac{a + b}{a - b} = \dfrac{c + d}{c - d}$ obtained from the proportion $a/b = c/d$ if $a \neq b$.

proportion by alternation The proportion $a/c = b/d$ obtained from the proportion $a/b = c/d$ if $c \neq 0$.

proportion by inversion The proportion $b/a = d/c$ obtained from the proportion $a/b = c/d$ if a and c are nonzero.

proportion by subtraction The proportion $\dfrac{a - b}{b} = \dfrac{c - d}{d}$ obtained from the proportion $a/b = c/d$.

proportional One of the terms in a proportion. *See,* for example, FOURTH PROPORTIONAL.

proposition *See* THEOREM.

prototile A region bounded by a SIMPLE CLOSED CURVE that is used as a tile in a tiling.

protractor An instrument used to measure angles.

pseudo-Euclidean geometry Coordinate geometry in which the distance between the points (x_1, y_1, z_1, t_1) and (x_2, y_2, z_2, t_2) is $\sqrt{(x_1 - x_2)^2 + (y_1 - y_2)^2 + (z_1 - z_2)^2 - (t_1 - t_2)^2}$.

pseudorhombicuboctahedron A NONUNIFORM POLYHEDRON with eight triangular faces and 18 square faces.

pseudosphere A model of the HYPERBOLIC PLANE that is embedded in Euclidean three-space. It is the surface formed by rotating one half of a TRACTRIX about its asymptote.

pseudovertex A point at the end of the MINOR AXIS of an ellipse.

purely imaginary number A complex number of the form ai where a is a real number.

pursuit curve The path of a particle as it "pursues" a particle moving along a given curve. The pursuit curve depends on the given curve and the initial positions of the two particles. For example, a TRACTRIX is a pursuit curve.

pyramid A polyhedron formed by connecting each vertex of a polygon, called the base of the pyramid, to a point, called the apex of the pyramid, that is not coplanar to the base. The triangular faces thus formed are called the lateral faces.

pyritohedron A polyhedron with 12 congruent faces, each an irregular pentagon with BILATERAL SYMMETRY. Pyrite crystals have this shape.

Pythagorean identities The trigonometric formulas derived from the Pythagorean theorem. *See* TRIGONOMETRIC FORMULAS in chapter 3.

Pythagorean spiral A spiral formed by attaching right triangles with legs 1 and $\sqrt{n}$ for $n = 1, 2, 3, \ldots$. The triangles in a Pythagorean spiral have a common vertex and the hypotenuse of the triangle with legs 1 and $\sqrt{n}$ is the leg of the next triangle.

Pythagorean triangle A triangle all of whose sides have integer length.

Pythagorean triple Three integers a, b, and c such that $a^2 + b^2 = c^2$. If a, b, and c form a Pythagorean triple, there is a right triangle with sides equal to a, b, and c.

QED *Quod erat demonstrandum;* Latin for "that which was to be proved." Written at the end of a proof, after the conclusion.

QEF *Quod erat faciendum;* Latin for "that which was to be done." Written at the end of a construction.

quadrangle A set of four points, no three of which are collinear.

Pseudosphere

quadrangular set The six points that are the intersections of the six sides of a COMPLETE QUADRANGLE with a line. In some cases, these six points may not be distinct.

quadrant (1) One-quarter of the Cartesian plane. In the first quadrant, the x- and y-coordinates are both positive; in the second quadrant, just the y-coordinate is positive; in the third quadrant, both coordinates are negative; and in the fourth quadrant, just the x-coordinate is positive. (2) An instrument for measuring altitudes or angular elevations. It is used in astronomy and navigation.

quadratic equation An equation of the form $ax^2 + bx + c = 0$.

quadratic form A HOMOGENEOUS polynomial of degree 2.

quadratic number A number, such as 2 or $2 + \sqrt{2}$, that is the root of a quadratic equation with rational coefficients. Segments whose lengths are quadratic numbers can be constructed with compass and straightedge.

quadratic surface *See* QUADRIC.

quadratrix A curve that can be used to SQUARE THE CIRCLE.

quadrature of the circle *See* SQUARING THE CIRCLE.

quadric A surface in three-dimensional space satisfying a second-degree equation. For example, the sphere is a quadric because it satisfies the second-degree equation $x^2 + y^2 + z^2 = r^2$, r constant.

quadrifigure *See* COMPLETE QUADRILATERAL.

quadrifolium A curve with four petals given in polar coordinates by the equation $r = \sin(2\theta)$.

quadrirectangular tetrahedron A space-filling tetrahedron with three perpendicular edges that share a common vertex. All four faces are right triangles.

quantum geometry *See* NONCOMMUTATIVE GEOMETRY.

quartic Of the fourth degree.

quartrefoil A MULTIFOIL drawn outside of a square.

quasicrystal A crystal that does not have TRANSLATIONAL SYMMETRY but does have fivefold rotational symmetry.

quasiregular Describing a polyhedron or tiling with two types of congruent faces or tiles such that a face or tile of one type is completely surrounded by a face or tile of the other type.

quasirhombicosidodecahedron A NONCONVEX UNIFORM polyhedron with 62 faces: 20 triangles, 30 squares, and 12 pentagrams.

quaternion A number of the form $a + bi + cj + dk$, where $i^2 = j^2 = k^2 = -1$ and $ij = -ij = k$, $ik = -ki = -j$, and $jk = -kj = i$.

queen One of the seven different VERTEX NEIGHBORHOODS that can occur in a PENROSE TILING.

query object In computer vision, an object whose nature is to be ascertained by a computer from specific measurements of the object.

quotient geometry For a given point in a PROJECTIVE GEOMETRY, the quotient geometry is the geometry whose points are all lines through the given point and whose lines are all planes through the given point.

quotient space *See* IDENTIFICATION SPACE.

rabattement A technique used to create perspective drawing in which the ground plane is drawn above the picture plane.

radial curve A curve that is determined from the CURVATURE of a given curve. A point on the radial curve is the tip of the radius vector of the circle of curvature at its point of contact displaced so that its tail is at the origin.

radial symmetry The symmetry that a circle has. An object with radial symmetry has rotational symmetry through any angle about its center.

radian A unit of angle measure; 2π radians = 360°. An angle of one radian subtends an arc of a circle with the same length as the radius of the circle.

radical axis The locus of all points whose POWERS with respect to two circles are equal.

radical center For three circles whose centers are not collinear, the radical center is the point that has the same POWER with respect to each circle.

radical circle A circle that is orthogonal to three circles and has its center at the RADICAL CENTER of the three circles.

radical plane The plane containing the circle of intersection of two intersecting spheres.

radiolaria Microscopic marine animals whose skeletons are often in the shape of regular polyhedra.

radius A segment with one endpoint at the center of a circle and the other endpoint located on the circle.

radius of a regular polygon A segment connecting the center of the polygon to any of its vertices.

radius of curvature The radius of a CIRCLE OF CURVATURE or a SPHERE OF CURVATURE.

radius of inversion The radius of a CIRCLE OF INVERSION.

radius of torsion For a point on a curve, the radius of torsion is $\left|\frac{1}{\tau}\right|$, where τ is the TORSION.

radius vector *See* POSITION VECTOR.

raft A connected set of TILES in which adjacent tiles share an edge.

ramphoid cusp *See* CUSP OF THE SECOND KIND.

range The set of all possible outputs of a function.

range of points A collection of two or more collinear points. In some contexts, a range of points is all the points on a line.

rank (1) The number of LINEARLY INDEPENDENT ROWS or COLUMNS of a MATRIX. It is equal to the dimension of the image of a LINEAR TRANSFORMATION represented by the matrix. (2) The number of elements in a MAXIMAL FLAG of a geometry; it is defined only if each maximal flag of the geometry has exactly the same number of elements.

ratio A quotient or the result of division. Thus, the ratio of 3 to 4 is 3/4 or .75. Sometimes the ratio of a to b is written $a{:}b$.

ratio of dilation For a given DILATION, the DIRECTED MAGNITUDE of OP'/OP where O is the center of dilation and P' is the image of some P under the dilation.

ratio of division For a point P in the interior of segment AB, the ratio AP/PB.

ratio of magnification *See* RATIO OF SIMILITUDE.

ratio of similitude The ratio of the length of any segment to the length of its PREIMAGE under a SIMILARITY TRANSFORMATION. It is the constant factor by which a similarity multiplies every distance.

rational conic A conic given by the equation $ax^2 + bxy + cy^2 + dx + f = 0$, where a, b, c, d, e, and f are rational numbers.

rational expression A fraction whose numerator and denominator are polynomials. The denominator must be nonzero.

rational geometry Any geometry in which every line is the HARMONIC NET of any three of its points.

rational line A line given by the equation $ax + by + c = 0$, where a, b, and c are rational numbers.

rational number A ratio of two integers. The decimal expansion of a rational number is a repeating or terminating decimal.

rational plane A plane that contains a two-dimensional SUBLATTICE of a three-dimensional LATTICE.

rational point A point in the Cartesian plane whose coordinates are rational numbers.

rational triangle A triangle all of whose sides have rational lengths.

ray All points on a line on one side of a point (called the endpoint of the ray) including the endpoint. A ray is sometimes called a closed ray; an open ray is a ray without its endpoint.

reachability region The region that can be accessed by the hand of a robot arm.

real analysis ANALYSIS, particularly for functions defined on the real numbers.

real axis The horizontal axis of the complex plane.

real number line A line with a given correspondence between points on the line and the real numbers. The length of a segment on the real number line can be determined by taking the arithmetic difference of the real numbers corresponding to its two endpoints.

real numbers The set of numbers that includes the positive and negative integers, the rationals, and the irrationals.

real part For a complex number $a + bi$, the real part is a.

real vector space A VECTOR SPACE whose field of scalar is the real numbers.

reciprocal transversals If three ISOTOMIC POINTS, one on each of the three sides of a triangle, are collinear, then so are the other three isotomic points. The two lines of collinearity are reciprocal transversals.

reciprocation Reciprocation is a POLARITY determined by a fixed circle. The image of a point is the line perpendicular to the radius of the circle through the INVERSE of the point with respect to the circle and the image of a line is the INVERSIVE IMAGE with respect to the circle of the foot of the perpendicular dropped from the center of the circle to the line. The reciprocal of a line is its pole and the reciprocal of a point is its polar. Reciprocation with respect to a sphere is defined similarly.

rectangle A parallelogram with four right angles.

rectangular axes Axes that are perpendicular to one another.

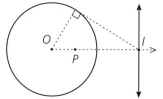

Point *P* and line *l* are reciprocal with respect to the circle with center *O*

rectangular coordinates Coordinates with respect to perpendicular axes.

rectangular hyperbola A hyperbola whose ASYMPTOTES are perpendicular to each other.

rectangular parallelepiped A PARALLELEPIPED with rectangular faces.

rectifiable curve A curve whose arc length can be determined.

rectifying plane For a point on a curve, the plane determined by the TANGENT and the BINORMAL. It is perpendicular to the NORMAL and OSCULATING PLANES.

rectilineal *See* RECTILINEAR.

rectilinear Describing a figure constructed from lines, segments, or rays.

recurrence relation An equation giving a term of a sequence as a function of earlier terms in the sequence. For example, $a_n = a_{n-1} + a_{n-2}$ is a recurrence relation that defines the FIBONACCI SEQUENCE.

recursion Repeated application of a function, transformation, or algorithm.

reduced generator A TRANSLATION of a tiling or LATTICE along a side of a REDUCED PARALLELOGRAM.

reduced parallelogram The FUNDAMENTAL REGION of a tiling or LATTICE that is a parallelogram with the shortest possible sides.

reductio ad absurdum *See* PROOF BY CONTRADICTION.

reduction A systematic translation from one theorem or problem to another, usually simpler, theorem or problem.

reentrant angle A REFLEX ANGLE of a NONCONVEX polygon.

reentrant polygon A polygon that is not CONVEX.

reference triangle A right triangle used to compute the values of the trigonometric functions of a directed angle in the Cartesian plane. One vertex is at the origin, one leg is on the x-axis, the other leg is parallel to the y-axis, and the hypotenuse is the terminal side of the angle.

reflection An isometry of the plane that leaves one line (the mirror line) fixed and maps each point to a point directly opposite on the other side of the mirror line. In three-dimensional space, a reflection is an isometry that leaves one plane (the mirror plane) fixed and maps each point to a point directly opposite on the other side the mirror plane.

Reflection

reflection in a point *See* CENTRAL INVERSION.

reflection with respect to a circle *See* INVERSION.

reflection, angle of The angle between the trajectory of a particle or ray of light reflecting off a surface and the NORMAL to the surface at the point of incidence.

reflex angle An angle with measure greater than 180°.

reflexive relation A RELATION with the property that any element is related to itself. On the integers, equality is a reflexive relation since $n = n$ for each integer n. However, less than is not a reflexive relation since $5 < 5$ is false.

region A set of points. In some contexts, a region is a connected set of points.

regular configuration A CONFIGURATION whose GROUP of AUTOMORPHISMS is TRANSITIVE on the points of the configuration.

regular conic projection The projection of a sphere onto a cone that is tangent to the sphere and whose axis runs through the poles of the sphere.

regular curve A curve all of whose points are regular points.

regular cylindrical projection A CYLINDRICAL PROJECTION of a sphere onto a cylinder that is tangent to the sphere along its equator.

regular graph A graph in which every vertex has the same degree.

regular point A point on a curve or surface where the tangent line or tangent plane is well defined and changes smoothly.

regular polygon A polygon in which all sides are congruent to one another and all angles are congruent to one another.

regular polyhedron A polyhedron in which all faces are congruent regular polygons and all vertices have the same degree.

regular polytope A POLYTOPE whose faces are regular polytopes that are congruent to one another.

regular simplex A SIMPLEX in which all the edges are congruent.

regular tiling A PERIODIC UNIFORM TILING by congruent regular polygons. There are only three such tilings: by equilateral triangles, by squares, and by regular hexagons.

regular vertex A vertex in a tiling for which all the incident angles are congruent to one another.

regulus The locus of lines intersecting three given SKEW LINES.

Reidemeister moves The transformations that change a knot diagram into an equivalent knot diagram. They are as follows:

1) a kink can be added or removed;
2) a strand can be slid over or under a looped strand; and
3) a strand can be slid over or under a crossing.

relation A way of comparing two elements of a set. For example, "less than" is a relation on the integers, and $5 < 6$ is written to show that 5 is less than 6. A relation can also be regarded as a collection of ordered pairs of elements of a set; if the relation is represented by the symbol R, $a\,\mathrm{R}\,b$ is written if the ordered pair (a, b) is in the relation.

remote angle Either of the two angles of a triangle that are not adjacent to a given EXTERIOR ANGLE.

rep-k tile A REPTILE that can be cut into exactly k congruent pieces. *See also* k-REP TILE.

repeller A point with the property that nearby points are mapped further and further away by a DYNAMICAL SYSTEM.

repelling fixed point A FIXED POINT of a DYNAMICAL SYSTEM that is a REPELLER.

reptile A polygon that can be cut into congruent polygons, each similar to the original polygon. For example, a square is a reptile because it can be cut into four congruent squares.

repulsive fixed point *See* REPELLING FIXED POINT.

respectively In the order mentioned. For example, the squares of 1, 2, and 3 are respectively 1, 4, and 9.

resultant The vector sum of two vectors.

Reuleaux polygon A region of constant width formed from a polygon with an odd number of sides.

Reuleaux triangle A region of constant width formed from an equilateral triangle.

reversal *See* INDIRECT ISOMETRY.

reverse of a knot The KNOT with the opposite ORIENTATION of the given knot.

reversible knot An ORIENTED KNOT that is equivalent to its REVERSE.

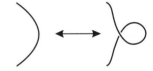

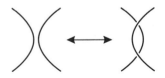

The three Reidemeister moves

Reye's configuration A CONFIGURATION in three-dimensional space consisting of 12 points and 12 planes. Each point is on six planes and each plane contains six points.

rhomb *See* LOZENGE.

rhombic disphenoid A polyhedron with four congruent scalene triangular faces and congruent solid angles.

rhombic dodecahedron A space-filling polyhedron with 12 faces, each a rhombus. It is the DUAL of the CUBOCTAHEDRON.

rhombic solid A polyhedron whose faces are congruent RHOMBS.

rhombic triacontahedron A ZONOHEDRON with 30 faces, each a RHOMBUS. It is the DUAL of the ICOSIDODECAHEDRON and has six ZONES.

rhombichexahedron *See* RHOMBOHEDRON.

rhombicosidodecahedron A SEMIREGULAR POLYHEDRON with 20 triangular faces, 30 square faces, and 12 pentagonal faces.

rhombicuboctahedron A SEMIREGULAR POLYHEDRON with eight triangular faces and 18 square faces.

rhombitruncated cuboctahedron A SEMIREGULAR POLYHEDRON with 12 square faces, eight hexagonal faces, and six octagonal faces.

rhombitruncated icosidodecahedron A SEMIREGULAR POLYHEDRON with 30 square faces, 20 hexagonal faces, and 12 decagon faces.

rhombohedral lattice A LATTICE whose LATTICE UNIT is a regular RHOMBOHEDRON.

rhombohedron A PARALLELEPIPED whose six faces are RHOMBS, or in some contexts, parallelograms.

rhomboid A parallelogram that is not a rectangle.

rhombus A parallelogram with all sides congruent.

rhumb line The course of a ship traveling with constant compass heading; *see* LOXODROME.

Riemann geometry The study of RIEMANN MANIFOLDS, specifically including the RIEMANN SPHERE and ELLIPTIC GEOMETRY.

Riemann manifold A MANIFOLD with complex structure. For example, the RIEMANN SPHERE is a Riemann manifold.

Riemann sphere The complex plane with an IDEAL POINT added.

right angle An angle with measure 90°.

right circular cone A cone whose base is perpendicular to its axis.

right circular cylinder A cylinder whose base is a circle orthogonal to the lateral surface.

right helicoid A HELICOID whose axis is perpendicular to its GENERATOR.

right prism A prism in which the lateral faces are perpendicular to the two bases.

right triangle A triangle with one right angle.

right-handed Having the same orientation as the thumb and first two fingers of the right hand held open with the thumb up, the index finger extended, and the middle finger bent.

rigid framework A FRAMEWORK that allows no MOTIONS other than congruences. For example, a triangle with a hinge at each vertex is rigid while a rectangle with a hinge at each vertex is not rigid.

rigid motion An ISOMETRY that preserves ORIENTATION.

rigor The use of precision and logic in mathematics.

ring *See* ANNULUS.

rise For two points in the Cartesian plane, the rise is the difference between their y-coordinates.

Robinson's tiles A set of six PROTOTILES that tile the plane only APERIODICALLY.

robot arm A sequence of segments, called links, connected at vertices, called joints, that are free to rotate. One endpoint, called the shoulder, is fixed and the other, called the hand, is moveable.

robot kinematics The study of the movement of the separate parts of a robot necessary to accomplish a specific task.

robotics The study of the design and control of robots designed to carry out specific tasks.

rod and joint framework A FRAMEWORK consisting of rods joined by flexible joints at their endpoints.

rod group *See* CYLINDRICAL GROUP.

Roman surface A NONORIENTABLE, one-sided, SIMPLE CLOSED SURFACE that cannot be embedded in three-dimensional space.

rosette A design with only DISCRETE ROTATIONAL or REFLECTIONAL SYMMETRIES.

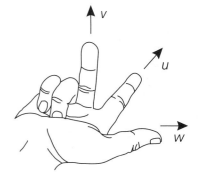

Right-hand system of vectors u, v, and w

rosette group The SYMMETRY GROUP of a finite design. A rosette group is either a CYCLIC GROUP or a DIHEDRAL GROUP.

rotary homography The product of two INVERSIONS with respect to intersecting circles.

rotary inversion *See* ROTARY REFLECTION.

rotary reflection An ISOMETRY of three-dimensional space that is the composite of a REFLECTION with a ROTATION whose axis is perpendicular to the mirror plane of the reflection.

rotation An ISOMETRY of the plane that leaves one point (the center of rotation) fixed and moves other points on a circle centered at the center of rotation through a fixed angle. Also, an isometry of space that leaves one line (the axis of rotation) fixed and moves other points on a circle centered on and orthogonal to the axis of rotation through a fixed angle.

rotation index The TOTAL CURVATURE of a CLOSED curve divided by 360°. It gives the number of full turns a unit tangent vector makes when the curve is traversed once counterclockwise.

rotational symmetry Symmetry about a point (called a center of symmetry) or line (called an axis of rotation).

rotocenter The center of a ROTATION; it is left fixed by the rotation.

rotoinversion A ROTATION followed by REFLECTION in a point that lies on the rotation axis.

rotoinversion axis The rotation axis of a ROTOINVERSION.

rotor A curve inscribed in a polygon that can rotate 360° and remain tangent to all sides of the polygon.

row A horizontal line of elements in a MATRIX.

row matrix A MATRIX with one row.

row vector *See* ROW MATRIX.

Rule of Three The rule that says that the product of the means of a proportion is equal to the product of the extremes. Thus, for fractions, if *a/b* = *c/d*, then *ad* = *bc*.

ruled surface A surface that can be generated by a specified motion of a line in three-dimensional space. For example, a cylinder is a ruled surface since it is generated by a line moving perpendicularly to the points on a given circle.

Rotation

ruler An instrument used to measure the distance between two points or the length of a segment.

Ruler Axiom The axiom that says that the points on a line can be put into one-to-one correspondence with the set of real numbers in such a way that the distance between two points is the absolute value of the difference of the corresponding real numbers.

ruler dimension *See* DIVIDER DIMENSION.

ruling A set of nonintersecting lines on a surface such that each point on the surface lies on exactly one of the lines.

run For two points in the Cartesian plane, the difference between their x-coordinates.

rusty compass A compass that can construct circles with a radius of one fixed length only.

SAA *See* AAS.

Saccheri quadrilateral A quadrilateral having two right angles and two congruent sides. The right angles are adjacent; the congruent sides are opposite each other; and each of the congruent sides is a leg of one of the right angles. It is used in the study of hyperbolic geometry.

saddle point (1) A HYPERBOLIC POINT on a surface. (2) A point in a vector field with the property that some nearby vectors point toward it and some nearby vectors point away from it.

saddle surface A surface that contains a HYPERBOLIC, or saddle, point.

salient angle An angle of a nonconvex polygon with measure less than 180°.

salient point A corner point of a curve where the TANGENTS on both sides of the point are well defined and different.

Salmon point A point of concurrency of four CAYLEY LINES. Any six points lying on a conic section determine 15 Salmon points.

SAS *See* SIDE-ANGLE-SIDE.

satisfiable Describing an AXIOMATIC SYSTEM that has an interpretation in which all the axioms are true.

scalar A number, especially in the context of linear algebra.

scalar curvature The absolute value of the CURVATURE.

scalar field An assignment of a SCALAR to each point in a region.

scalar function A function from a VECTOR SPACE to its field of scalars.

scalar multiple A VECTOR that has been multiplied by a scalar.

scalar multiplication The multiplication of a VECTOR by a scalar. The direction of the new vector is the same (or opposite if the scalar is negative) and its magnitude is the original magnitude multiplied by the scalar. In coordinates, the vector (a, b) multiplied by the scalar s is (sa, sb).

scalar product *See* DOT PRODUCT.

scalar torsion The length of a TORSION VECTOR.

scalar triple product For VECTORS $\mathbf{u}$, $\mathbf{v}$, and $\mathbf{w}$, the vector $\mathbf{u} \cdot (\mathbf{v} \times \mathbf{w})$, denoted $(\mathbf{u} \, \mathbf{v} \, \mathbf{w})$. The magnitude of the scalar triple product is equal to the parallelepiped determined by the three vectors.

scale factor *See* RATIO OF SIMILITUDE.

scalene triangle A triangle with no two sides or angles congruent.

scaling factor *See* RATIO OF SIMILITUDE.

scan conversion The process of determining the pixels on a computer screen that correspond to a given geometric figure.

scaphe An ancient instrument used to determine the INCLINATION of the sun. It consists of a hemispherical bowl and gnomon which casts a shadow inside the bowl.

scenic path A path between two vertices of a VISIBILITY GRAPH that is completely visible from a given viewpoint.

Schläfli symbol A symbol (p, q) or $\{p, q\}$ used to represent a regular polyhedron (or tiling) whose faces (tiles) are regular p-gons and whose vertices have degree q.

Schläfli's double-six A CONFIGURATION in three-dimensional space consisting of 30 points and 12 ("double-six") lines. Each point is on two lines and each line contains five points.

Schlegel diagram A planar representation of a polyhedron obtained by cutting away the interior of one face and then flattening the remaining surface without tearing it. The punctured face corresponds to an infinite region in the plane.

Schmitt-Conway biprism A CONVEX polyhedron that fills three-dimensional space only APERIODICALLY.

Schwarz inequality The statement that the absolute value of the DOT PRODUCT of two vectors is less than or equal to the product of their magnitudes. Thus, $|\mathbf{v}\cdot\mathbf{w}| \leq \|\mathbf{v}\| \|\mathbf{w}\|$ for vectors $\mathbf{v}$ and $\mathbf{w}$.

screw displacement *See* TWIST.

screw rotation *See* TWIST.

scroll A ruled surface that is not developable.

sec *See* SECANT.

secant For an acute angle in a right triangle, the ratio of the hypotenuse over the adjacent side.

secant line A line intersecting a circle or other curve in two or more points.

second One 60th of a minute; a unit of angle measure.

second Brocard triangle The triangle inscribed in the BROCARD CIRCLE of a given triangle such that each vertex is the intersection of the Brocard circle with a ray from a vertex of the triangle to the SYMMEDIAN POINT of the triangle.

second curvature The second curvature of a curve is its TORSION.

second Lemoine circle For each side of a triangle, construct the ANTIPARALLEL, with respect to the other two sides, that passes through the SYMMEDIAN POINT. The second Lemoine circle passes through the six points of intersection of these antiparallels with the sides of the triangle.

second-order cone A surface formed by connecting each point on a conic section to a fixed point not on the plane of the conic section.

section The intersection of a line or plane with some other set.

sector The region of a circle bounded by a central angle and the intercepted arc.

sec⁻¹ The INVERSE of the secant function.

seed *See* VORONOI DIAGRAM.

segment (1) All points between and including two points (called the endpoints of the segment) on a line. (2) The region in a circle between a chord of the circle and the circumference of the circle.

Seifert surface An ORIENTABLE surface for which a given KNOT is the boundary.

self-congruent (1) Describing a shape that has an ISOMETRY, different from the identity, mapping the shape to itself. (2) Describing an object that is conjugate to itself or is invariant under a conjugacy operation.

self-conjugate line A line that contains its image under a POLARITY.

self-conjugate point A point that lies on its image under a POLARITY.

self-conjugate triangle *See* SELF-POLAR TRIANGLE.

self-dual Describing an object, such as a polyhedron or tiling, that is dual to itself.

self-dual configuration A CONFIGURATION in which the number of points is equal to the number of lines.

self-inverse A point on a CIRCLE OF INVERSION; so called because such a point is its own inverse.

self-polar configuration A CONFIGURATION that admits a POLARITY.

self-polar triangle A triangle in which each vertex is the POLE of the opposite side.

self-similar A self-similar shape has the property that parts of the shape are similar to other parts or to the whole. Fractals are self-similar.

self-similar dimension The dimension of a SELF-SIMILAR set defined by $\dfrac{\log N}{\log(1/s)}$, where the fractal consists of N pieces that are scaled by a factor of s. Thus, since the CANTOR SET consists of two copies of itself scaled by a factor of 1/3, its fractal dimension is $\dfrac{\log 2}{\log 3} \approx .6309$.

semidiagonal A segment joining the midpoints of two opposite edges of a cube. It passes through the center of the cube and is an axis of rotational symmetry of the cube.

semi-Euclidean geometry Coordinate geometry in which the distance between two points is the absolute value of the difference of the first coordinates of the two points.

semi-major axis Half of the MAJOR AXIS of an ellipse.

semi-minor axis Half of the MINOR AXIS of an ellipse.

semicircle Half of a circle.

semicubical parabola The curve given by the equation $ay^2 = x^3$.

semilune A triangle bounded by parts of three great circles that has two right angles. It is half of a lune.

semiperimeter Half of the perimeter of a polygon.

semiregular polyhedron A UNIFORM POLYHEDRON whose faces are REGULAR POLYGONS, but are not all congruent. There are 13 different semiregular polyhedra.

semiregular polytope A POLYTOPE whose faces are REGULAR, but not all congruent, polytopes.

semiregular tiling A PERIODIC UNIFORM TILING by REGULAR POLYGONS that are not all congruent. There are eight semiregular tilings.

semitangent A boundary ray of a TANGENT CONE.

sense *See* ORIENTATION. *Sensed* means "oriented" and *sense-preserving* means "orientation-preserving."

sensed magnitude *See* DIRECTED MAGNITUDE.

sensed parallel In HYPERBOLIC GEOMETRY, a parallel to a given line through a given point together with a direction, such that a right-sensed or right-hand parallel will intersect the given line if it is rotated any amount clockwise about the given point and a left-sensed or left-hand parallel will intersect the given line if it is rotated any amount counterclockwise about the given point.

sensed segment *See* DIRECTED SEGMENT.

separate (1) In general, to come between; to divide into two or more parts. (2) A pair of points A and B separate another pair of points C and D if the cross ratio (AB, CD) is negative.

separatrix A curve that separates regions. For example, a separatrix in a PHASE PORTRAIT separates BASINS.

sequence A list of numbers $a_1, a_2, a_3, a_4, \ldots$. An infinite sequence of numbers converges if there is a number, called the limit of the sequence, such that the numbers in the sequence keep getting closer and closer to the limit.

series A sum of a sequence of numbers, $a_1 + a_2 + a_3 + \ldots$. An infinite sequence converges if the sequence of its partial sums $a_1, a_1 + a_2, a_1 + a_2 + a_3, \ldots$ converges.

serpentine curve The S-shaped curve defined by the equation $x^2 y + b^2 y - a^2 x = 0$ for a and b constant. It is asymptotic to the x-axis in both directions.

set A collection of definite, distinct objects, which are called the elements of the set. Usually, set is an undefined term. Sets are

The eight semiregular tilings

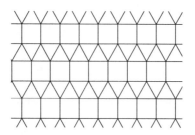
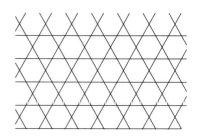
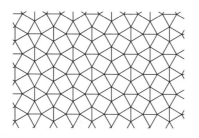
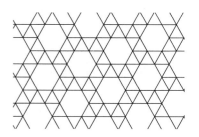
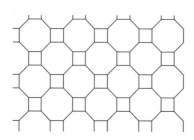
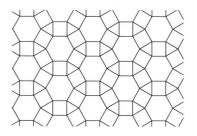
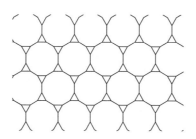
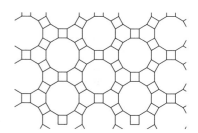

denoted using braces; thus $\{a, b, c\}$ is the set containing the elements a, b, and c. *See also* ELEMENT.

set difference The elements of one set that are not contained in another. The set difference $S \backslash T$ is the set of all elements of S that are not elements of T.

set theory The mathematical theory that describes collections of objects (sets) and how they are related to their elements and to one another. *See also* SET.

setting-out A description of the conditions and hypotheses for a theorem.

sextant A mechanical device used to measure the angle between two distinct objects, used mainly in navigation to determine longitude.

shadow segment A segment on a VISIBILITY MAP that separates an invisible region closer to the viewpoint from a visible region further from the viewpoint.

sheaf A set of planes through a point.

shear An EQUIAFFINE TRANSFORMATION that leaves a given line pointwise fixed and maps lines parallel to the given line to themselves.

sheet A connected surface.

shoemaker's knife *See* ARBELOS.

shortest-route algorithm An algorithm for finding the shortest path between two points on a surface or between two vertices on a GRAPH.

shortest-watchtower algorithm An algorithm that determines the lowest elevation above a TRIANGULATED TERRAIN from which the entire terrain is visible.

shoulder *See* ROBOT ARM.

shrink A transformation of the plane given by the equations $f(x, y) = (kx, ky)$ where $0 < k < 1$.

Siamese dodecahedron *See* SNUB DISPHENOID.

side (1) In a polygon, a segment connecting two adjacent vertices. (2) One of the two rays or segments determining an angle.

side-angle-side If two sides of one triangle are congruent or proportional to the sides of another triangle and the included angles are congruent, the triangles are congruent or similar, respectively.

side-side-side If three sides of one triangle are congruent or proportional to the sides of another triangle, the triangles are congruent or similar, respectively.

sideline The line determined by a side of a polygon.

Sierpinski carpet A fractal obtained by dividing a square into nine congruent squares and removing the middle square and then iterating this procedure on each of the remaining squares.

Sierpinski gasket *See* SIERPINSKI SPONGE.

Sierpinski sieve *See* SIERPINSKI CARPET.

Sierpinski space A TOPOLOGICAL SPACE containing two points and three open sets.

Sierpinski space-filling curve A fractal curve that has infinite length, encloses a finite area, and passes through every point of the unit square.

Sierpinski sponge A fractal obtained by dividing a cube into 27 congruent cubes and removing seven cubes—the central cube and the six cubes that share a face with it—and then iterating this procedure on each of the remaining cubes.

Sierpinski triangle A fractal obtained by dividing an equilateral triangle into four congruent equilateral triangles and removing the middle triangle and then iterating this procedure on each of the remaining triangles.

sighting angle *See* ANGLE OF SIGHT.

sign of a permutation The sign of a permutation is +1 if the permutation is even and –1 if it is odd.

signed magnitude A measurement with a positive or negative sign associated with it depending on the orientation of the object measured. For example, the directed magnitude of the length of segment AB is the negative of that of segment BA and the directed magnitude of the measure of a clockwise angle is negative.

signum function The function sign (x) that is defined to be 1 for $x < 0$, 0 for $x = 0$, and –1 for $x > 1$.

similar Having the same shape but not necessarily the same size. Angular measurements made in similar figures are equal and linear measurements made in similar figures are proportional.

similar polygons Two polygons that are SIMILAR. There is a one-to-one correspondence between angles and sides with corresponding angles congruent and corresponding sides proportional.

similarity A transformation from a space to itself that multiplies every distance by the same number.

similarity tiling A tiling that is equal to a composition of itself. A similarity is called a *k*-similarity tiling if the tiles in the composition are the unions of exactly *k* of the original tiles.

similitude *See* SIMILARITY.

similitude, center of For two circles, the center of a DILATION that takes one of them to the other.

simple circuit A CIRCUIT in a GRAPH that does not pass through any vertex more than once.

simple close packing A PACKING of three-dimensional space by spheres in which the centers of the spheres are at the points of a LATTICE formed from cubes. Each sphere is tangent to six other spheres.

simple curve A curve that does not intersect itself.

simple point *See* REGULAR POINT.

simple polygon A polygon that does not intersect itself.

simple polyhedron A polyhedron that can be deformed into a sphere.

simplex An *n*-dimensional simplex is the convex hull of $n + 1$ points in general position in a space of dimension *n* or higher.

simplicial complex A collection of SIMPLICES that intersect only at lower-dimensional faces.

simplicial mapping A map between SIMPLICIAL COMPLEXES that takes simplices to simplices.

simplicity of a construction The total number of operations made in completing a particular construction.

simply connected A region is simply connected if every SIMPLE CLOSED CURVE in the region can be shrunk to a point while remaining in the region.

Simson line The line passing through the feet of the perpendiculars dropped from a given point on the circumcircle of a triangle to the three sides of the triangle. The given point is called the pole of the Simson line.

sin   *See* SINE.

sin⁻¹   *See* INVERSE SINE.

sine   For an acute angle in a right triangle, the ratio of the opposite side over the hypotenuse. A directed angle with vertex at the origin of the Cartesian plane determines a point (x, y) on the unit circle centered at the origin; the sine of the directed angle is y.

single elliptic geometry   An ELLIPTIC GEOMETRY in which two lines intersect at a unique point.

singular matrix   A square matrix whose DETERMINANT is 0.

singular point   A point on a curve that does not have a well-defined TANGENT LINE.

singular point of a tiling   A point in a tiling with the property that every disc centered at that point intersects infinitely many tiles.

sinh   *See* HYPERBOLIC SINE.

sink   A point in a vector field with the property that all nearby vectors point toward it.

site   *See* VORONOI DIAGRAM.

skeleton of a polygon   *See* MEDIAL AXIS.

skew lines   Lines in space that are not parallel but that also do not intersect.

skew polygon   A polygon whose vertices are not planar.

skew-symmetric matrix   A square matrix that is equal to the negative of its TRANSPOSE.

slant height   The length of a segment from the vertex to a point on the circumference of the base of a right circular cone or to the midpoint of an edge of the base of a regular pyramid.

slice knot   A KNOT that is the boundary of a disc in four-dimensional space.

sliding vector   A vector whose initial point can be any point on a given line parallel to the vector.

slope   For a line in the Cartesian plane, the rise over the run, computed for any two points on the line. The slope of the line passing through points (x_1, x_2) and (y_1, y_2) is $\frac{y_1 - y_2}{x_1 - x_2}$.

slope line   A curve that crosses the level curves of surface at right angles. An ascending slope line is a directed slope line pointing in the direction of increasing values of the level curves, and a descending slope line is a directed slope line pointing in the direction of decreasing values of the level curves.

small circle A circle on a sphere whose center does not coincide with the center of the sphere.

small cubicuboctahedron A NONCONVEX UNIFORM POLYHEDRON with eight triangular faces, six square faces, and six octagon faces. It is a faceted RHOMBICUBOCTAHEDRON.

small rhombicosidodecahedron *See* RHOMBICOSIDODECAHEDRON.

small rhombicuboctahedron A SEMIREGULAR POLYHEDRON with eight triangular faces and 18 square faces.

small stellated dodecahedron A dodecahedron STELLATED with pentagonal pyramids on each face. It has 60 triangular faces and can be formed by 12 pentagrams that pass through one another meeting five at each vertex. It is a KEPLER-POINSOT SOLID.

small stellated triacontahedron A ZONOHEDRON that is the dual of the dodecadodecahedron.

smooth Describing a curve or surface whose tangent lines or planes change continuously as one moves from point to point. For example, a circle is smooth while a square is not.

smooth function A function whose DERIVATIVES are continuous.

smuggler's path A path between two vertices of a visibility graph that is entirely invisible from a given viewpoint.

snake A HEXIAMOND shaped like a snake.

snowflake curve *See* KOCH SNOWFLAKE.

snub cube A SEMIREGULAR POLYHEDRON with 32 triangular faces and six square faces. It can be inscribed in a cube so that its square faces lie on the faces of the cube.

snub cuboctahedron *See* SNUB CUBE.

snub disphenoid A CONVEX polyhedron with 12 faces, each an equilateral triangle.

snub dodecahedron A SEMIREGULAR POLYHEDRON with 80 triangular faces and 12 pentagonal faces. It can be inscribed in a dodecagon so that its pentagonal faces lie on the faces of the dodecagon.

snub icosidodecahedron *See* SNUB DODECAHEDRON.

Soddy's circles Given any three points, three mutually tangent circles can be drawn with these points as center. Soddy's circles are the two circles that are tangent to all three such circles.

solid angle A three-dimensional figure formed by three or more noncoplanar rays with a common endpoint. As with angles, segments incident to the rays of a solid and to the vertex may be used to represent the angle. The measure of a solid angle given in STERRADIANS is determined by the part of the surface of a sphere cut out when the vertex of the angle is at the center of the sphere.

solid geometry The Euclidean geometry of three-dimensional space.

Soma cube A POLYCUBE made from three or four cubes. There are seven different Soma cubes.

source A point in a vector field with the property that all nearby vectors point away from it.

space The set of points belonging to a geometry. Thus, the points in a plane constitute Euclidean space and the points of a PROJECTIVE GEOMETRY constitute a projective space.

space diagonal A segment that joins two vertices of a polyhedron but does not lie on a face of the polyhedron.

space dual The dual of a statement in projective geometry obtained by replacing each occurrence of the word *point* by *plane* and each occurrence of the word *plane* by the word *point.*

space group (1) A GROUP of SYMMETRIES that contains translations in one or more directions. (2) *See* CRYSTALLOGRAPHIC GROUP.

space homology A transformation of three-dimensional space that is the composition of a DILATION with a rotation about an axis passing through the center of dilation.

space-filling curve A fractal curve that passes through every point in a two-dimensional region of space.

space-filling regions A set of regions in three-dimensional space with the property that each point in space is contained within one of the regions or on the boundary between regions.

Spaceland The term used for three-dimensional space by British clergyman E. A. ABBOTT in the satire *Flatland.*

spacetime geometry An ANALYTIC GEOMETRY used in physics in which one of the directions models time rather than space.

spacetime interval The distance between two points in a SPACETIME GEOMETRY.

span The set of all linear combinations of a given set of vectors.

species, given in A RECTILINEAR figure is given in species if the angles and ratios of the sides are given.

species of a vertex The arrangement of tiles about a vertex in a tiling by regular polygons. There are 17 different possible species.

specification *See* DEFINITION.

spectral radius The largest of the absolute values of the EIGENVALUES of a given matrix.

spectrum The set of EIGENVALUES of a given matrix.

speed Change in position with respect to time; speed is always given as a nonnegative real number.

sphere The set of all points in three-dimensional space at a given distance from a given point, which is called the center of the sphere. More generally, the term *sphere* is used for a HYPERSPHERE.

sphere of Apollonius The sphere of points whose distances from two fixed points are in a constant ratio.

sphere packing A PACKING by spheres.

spheric The name used by BUCKMINSTER FULLER for the rhombic dodecahedron.

spherical coordinates Coordinates for three-dimensional space with respect to two fixed axes meeting at the origin. A point has coordinates (r, θ, φ) where r is the distance of the point from the origin, θ gives the directed angle from a fixed horizontal axis to the ray from the origin to the point, and φ gives the directed angle from a fixed vertical axis to the ray from the origin to the point.

spherical curve A curve lying on the surface of a sphere.

spherical degree A measure for solid angles; one spherical degree is equal to $\pi/180$ STERRADIAN.

spherical excess *See* ANGULAR EXCESS.

spherical geometry The study of the geometry of the surface of a sphere.

spherical helix A helix that lies on a sphere.

Sphere

spherical layer The part of a ball between two parallel planes intersecting the ball.

spherical lune *See* DIGON.

spherical polygon A region of a sphere whose boundaries are arcs of great circles of the sphere.

spherical representation The GAUSS MAP from a surface to the unit sphere.

spherical segment The part of a ball on one side of a plane that intersects the ball.

spherical triangle A geometric figure on the surface of a sphere that has three vertices connected by portions of great circles.

spherical trigonometry The study of SPHERICAL TRIANGLES and the relationships among their measurements.

spherical wedge One of the four parts of a ball between two planes passing through a common diameter of the ball.

spherical zone The region on a sphere between two parallel planes intersecting the sphere.

spheroid *See* ELLIPSOID OF REVOLUTION.

sphinx A HEXIAMOND shaped like a sphinx. It is a REPTILE and can be cut into four congruent copies of itself.

spider The figure formed by a vertex of a tiling and its incident edges.

Spiecker circle The incircle of the MEDIAL TRIANGLE of a given triangle.

spinode *See* CUSP.

spira aurea *See* GOLDEN SPIRAL.

spira mirabilis The name given to the LOGARITHMIC SPIRAL by JACOB BERNOULLI.

spiral A planar curve traced out by a point rotating about a fixed point while simultaneously moving away from the fixed point.

spiral of Archimedes *See* ARCHIMEDEAN SPIRAL.

spiral of Cornu An S-shaped double spiral.

spiral rotation *See* DILATIVE ROTATION.

spiral similarity *See* DILATIVE ROTATION.

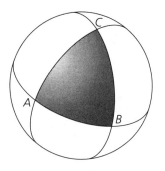

Spherical triangle *ABC*

spiric curve A section of a SPIRIC SURFACE by a plane parallel to the axis of rotation.

spiric surface A surface generated by a circle rotating about an axis in the plane of the circle. The TORUS is an example of a spiric surface.

spirograph An instrument for making patterns that have rotational symmetry.

spline A smooth curve passing through specified points in the plane.

splitter A segment starting at vertex of a triangle that bisects the perimeter of the triangle.

square A quadrilateral with all sides congruent to one other and all angles congruent to one other.

square matrix A matrix that has the same number of rows as columns.

squared rectangle A rectangle that has been dissected into squares of various sizes.

squaring the circle The problem of constructing a square with the same area as a given circle. One of the three famous problems of antiquity, it is impossible to solve with compass and straightedge, but can be solved, for example, using the ARCHIMEDEAN SPIRAL or the CONCHOID OF NICOMEDES.

SSS *See* SIDE-SIDE-SIDE.

stability domain The values of the PARAMETERS of a DYNAMICAL SYSTEM for which the EQUILIBRIUM STATE is STABLE.

stabilizer The elements of a SYMMETRY GROUP that leave a particular point or set fixed.

stabilizer of a tile *See* INDUCED TILE GROUP.

stable Unchanged or changed very little by small changes in parameters or input.

stage *n* fractal A PREFRACTAL that is obtained after *n* iterations of a process that will ultimately create a fractal.

stagnation point A point in a flow where the velocity is zero.

standard lattice The LATTICE containing all points in the Cartesian plane whose coordinates are integers.

star (1) A set of segments with a common midpoint. A star is nonsingular if no three of the segments are coplanar. (2) One of the seven different vertex neighborhoods that can occur in a Penrose tiling. (3) *See* BUNDLE.

star of Lakshmi *See* OCTAGRAM.

star polygon A NONCONVEX polygon formed by extending the sides of a given polygon until they intersect. For example, the PENTAGRAM is a star polygon.

starfish A type of DECAPOD.

state space A vector space used in physics in which each vector represents a state of a physical system. The components of a vector correspond to measurements of the system.

station point The location of the eye of the viewer in a PERSPECTIVE PROJECTION. It is the center of projection.

Steiner chain A finite TRAIN of circles in which each circle is tangent to two other circles in the train.

Steiner point (1) A point of intersection of three PASCAL LINES. Six points lying on a conic section determine 60 Pascal lines and thus 20 Steiner points. (2) The point of concurrency of three lines, each passing through a vertex of a triangle and parallel to a corresponding side of the first BROCARD TRIANGLE of the triangle. It is the center of mass of the system formed by putting a mass at each vertex of a triangle that is proportional to the measure of the EXTERIOR ANGLE at that vertex and it is on the circumcircle of the triangle.

Steiner symmetrization A way to transform a convex region in the plane into a symmetric region having the same area. Slice the given region into thin strips using parallel chords; draw an axis perpendicular to the chords; and slide each chord so that the axis divides it in half. The resulting shape will have bilateral symmetry.

Steiner tree The minimum-weight connected subgraph of a weighted graph that includes all the vertices.

stella octangula A STELLATED octahedron.

stellate To extend each of the faces of a polyhedron until they intersect one another.

stellated polyhedron A polyhedron formed by stellating each face of a polyhedron.

steregon A solid angle with measure equal to 720 SPHERICAL DEGREES or 4π STERRADIANS. It is the solid analog of a 360° angle in the plane.

stereochemical topology The study of the topology of the arrangement of atoms in a molecule.

stereographic projection A PERSPECTIVE PROJECTION of the surface of a sphere onto a plane. In stereographic projection, the image plane is tangent to the sphere and the center of projection is the point on the sphere antipodal to the point of tangency. Stereographic projection is CONFORMAL.

stereometry The science of determining the measurements of solid figures.

sterradian A unit of measure of solid angles, abbreviated sr. An angle of measure 1 sr intercepts an area of r^2 square units on the surface of a sphere of radius r.

stircle A figure that is either a line or a circle.

stochastic Random.

straight Having CURVATURE equal to 0.

straight angle An angle with measure equal to 180°.

straight line *See* LINE. The term *straight line* was used by Euclid for what is now called a line; Euclid's term *line* included what we now call curves.

straight object A line, ray, or segment.

straightedge An instrument used to construct a line in Euclidean geometry.

strain An AFFINE TRANSFORMATION that leaves a given line pointwise fixed and maps lines perpendicular to the given line to themselves.

strange attractor A set of ATTRACTORS of a DYNAMICAL SYSTEM that has an unexpected or chaotic appearance.

stretch *See* EXPANSION.

stretch-reflection *See* DILATIVE REFLECTION.

stretch-rotation *See* DILATIVE ROTATION.

stretching factor *See* RATIO OF SIMILITUDE.

strictly convex polyhedron A CONVEX polyhedron with the property that for each vertex there is a plane through that vertex that does not intersect the polyhedron at any other point.

string theory A foundational theory in physics in which particles are represented by one-dimensional vibrating loops or arcs.

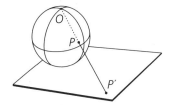

Stereographic projection with center of projection O

strip group   The GROUP OF SYMMETRIES of a band ornament or strip pattern. There are only seven possible strip groups.

strip pattern   *See* BAND ORNAMENT.

strip tiling   *See* FRIEZE GROUP.

strombic hexecontahedron   *See* TRAPEZOIDAL HEXECONTAHEDRON.

strophoid   A strophoid is the locus of points on a ray from a fixed point to a given curve that are equidistant from another fixed point. If the curve is a line and the two points are on a ray perpendicular to the line, the strophoid is called a right strophoid and has one loop and is asymptotic to the line in two directions.

sublattice   A LATTICE that is contained in some other lattice.

submatrix   A new matrix derived from a given matrix by deleting some rows or columns.

subset   A set, all of whose elements are also elements of another set. The notation $S \subseteq T$ means that S is a subset of T.

subspace topology   A TOPOLOGY given to a subset of a topological space. An open set in the subspace topology is the intersection of the subset with an open set in the original space.

subtend   To be located opposite something. For example, an arc of a circle subtends the central angle that contains it.

sufficient condition   The premise of an implication.

summit   The side opposite the base of a SACCHERI QUADRILATERAL.

summit angle   An angle with one leg incident to the SUMMIT of a Saccheri quadrilateral.

sun   One of the seven different VERTEX NEIGHBORHOODS that can occur in a PENROSE TILING.

superposition   The placing of one geometric figure on top of another.

supersymmetry   A symmetry between elementary particles in theoretical physics.

supplement of an angle   An angle whose measure added to the measure of the given angle gives 180°.

supplementary angles   Two angles whose measures add up to 180°.

supporting half-plane   The half-plane determined by a SUPPORTING LINE that contains the supported set.

supporting half-space The half-space determined by a SUPPORTING PLANE that contains the supported set.

supporting line A line passing through at least one boundary point of a planar set such that all points of the set are on the same side of the line.

supporting plane A plane through at least one boundary point of a set in three-dimensional space such that the set is completely contained in one of the closed half-spaces determined by the plane.

supremum For a set of numbers, the smallest number that is larger than every element of the set.

surface area A measure of the size of a surface or a region on a surface.

surface coordinates Coordinates with respect to a chart on some portion of a surface.

surface distance *See* INTRINSIC DISTANCE.

surface of centers The EVOLUTE of a surface.

surface of revolution A surface swept out by revolving a curve about an axis. For example, a cylinder is the surface of revolution obtained when a line is revolved about a line that is parallel to it.

surface of second order *See* QUADRIC SURFACE.

surgery The process of cutting out a piece of a surface and pasting in another surface with the same boundary as the piece that was removed. The new piece is attached along the boundary of the removed piece.

surjection *See* ONTO FUNCTION.

surjective function *See* ONTO FUNCTION.

swallowtail A type of three-dimensional CATASTROPHE or singularity that has the shape of a swallowtail.

symmedian The ISOGONAL CONJUGATE of a median of a triangle.

symmedian point The point of intersection of the three SYMMEDIANS of a triangle.

symmetric axis *See* MEDIAL AXIS.

symmetric configuration A CONFIGURATION in which the number of points is the same as the number of lines.

symmetric difference For two sets, the symmetric difference is the set that contains all elements in one, but not both, of the sets.

symmetric group The GROUP of all possible permutations or rearrangements of the numbers 1, 2, 3, . . ., n. It is denoted S_n.

symmetric matrix A square matrix that is equal to its TRANSPOSE.

symmetric points A pair of points such that one is the INVERSE POINT of the other.

symmetric relation A RELATION that is true if its order is reversed. On the integers, equality is a symmetric relation because if $a = b$, then $b = a$. However, "less than" is not a symmetric relation because $5 < 6$ is true, but $6 < 5$ is not true.

symmetrical Describing a geometric object or pattern that has nontrivial SYMMETRY TRANSFORMATIONS. Thus, a square is symmetrical but a scalene triangle is not.

symmetry Geometric repetition, balance, harmony, or sameness of structure. Also, a SYMMETRY TRANSFORMATION is often called a symmetry.

symmetry element An object used to define a specific SYMMETRY TRANSFORMATION. For example, a center of rotation, a mirror line, and an axis of rotation are symmetry elements.

symmetry group The GROUP of all possible SYMMETRY TRANSFORMATIONS of a given pattern or design. It is trivial if it consists of just the identity transformation.

symmetry operation *See* SYMMETRY TRANSFORMATION.

symmetry transformation A TRANSFORMATION that maps a geometric shape or design to itself, leaving it apparently unchanged.

symmetry type Patterns are of the same symmetry type if their SYMMETRY GROUPS are ISOMORPHIC.

synclastic An oval-shaped CLOSED surface.

synthetic geometry Geometry that does not use numerical coordinates to represent points.

synthetic proof A proof based on axioms and constructions rather than on numerical or algebraic computations.

system of pointers POINTERS located at the points of a LATTICE.

system of equations Two or more equations. A solution of a system of equations is a solution for each equation in the system.

τ *See* GOLDEN RATIO.

tac-locus The TAC-POINTS of a family of curves.

tac-point A point of intersection of two distinct curves of a family of curves where the curves have a common tangent line.

tacnode *See* POINT OF OSCULATION.

tail *See* INITIAL POINT.

tail wind A wind blowing in the same direction as a velocity vector.

tame knot A KNOT whose KNOT DIAGRAM has finitely many CROSSINGS.

tan *See* TANGENT.

tan⁻¹ *See* INVERSE TANGENT.

tangent For an acute angle in a right triangle, the ratio of the opposite side over the adjacent side. A directed angle with vertex at the origin of the Cartesian plane determines a point (x, y) on the unit circle centered at the origin; the tangent of the directed angle is y/x.

tangent A tangent to a point conic is a line in the plane of the point conic having exactly one point in common with the conic.

tangent bundle The collection of all lines tangent to a given curve or all planes tangent to a given surface.

tangent circles Two circles that intersect at only one point. At the point of intersection, they have a common tangent.

tangent cone The set of all rays from a given point on the boundary of a convex set that intersect the set or its boundary at a point different from the given point.

tangent indicatrix *See* TANGENTIAL INDICATRIX.

tangent line The line that most closely approximates a curve at a point. For example, on a circle, the tangent line at a point is perpendicular to the radius at that point.

tangent plane The plane that most closely approximates a surface at a point. For example, on a sphere, the tangent plane at a point is perpendicular to the radius at that point.

tangent vector A vector lying on a tangent line or tangent plane.

tangential image *See* TANGENTIAL INDICATRIX.

tangential indicatrix The image of a curve under the GAUSS MAP.

tangle Two disjoint arcs contained in a sphere that have their endpoints attached to the sphere, together with finitely many disjoint simple

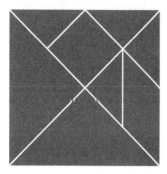

Tangrams

closed curves inside the sphere. Tangles are studied using the techniques of KNOT theory.

tangrams Seven pieces—five right triangles, one square, and one parallelogram—that together form a square. To complete a tangram puzzle, the seven pieces must be put together to form a given shape.

tanh *See* HYPERBOLIC TANGENT.

Tarry point The point on the CIRCUMCIRCLE of a triangle that is diametrically opposite the STEINER POINT of the triangle.

tautochrone A curve with the property that a particle subject to a specific force moving along the curve from any point to a specific point will take the same time. A CYCLOID is the tautochrone for a gravitational field.

taxicab metric A way to measure the distance between two points in the plane according to the path a taxicab would take. For points (a, b) and (c, d), the taxicab distance is $|a - c| + |b - d|$.

Taylor circle Drop perpendiculars from the feet of the altitudes of a given triangle to each of the opposite sides. The Taylor circle passes through the six feet of these perpendiculars.

Taylor series The POWER SERIES that most closely approximates a given function at a given point.

temple of Viviani The intersection of a sphere of radius $2r$ centered at the origin of three-dimensional space and a cylinder or radius r with axis passing through the point $(a, 0, 0)$ and parallel to the z-axis.

tensegrity framework A system of rods and cables held in place by tension in the cables and compression in the rods.

tensor Tensors represent a generalization of the concept of vector. A tensor consists of a sequence of vectors and scalar-valued functions defined on vector spaces. The tensor product of $\mathbf{v}$ and $\mathbf{w}$ is denoted $\mathbf{v} \otimes \mathbf{w}$.

term One expression that is added as part of a sum.

terminal point (1) The ending point of a vector. It is the point of the arrow representing the vector. (2) The final point of a directed curve.

terminal side *See* DIRECTED ANGLE.

terrain A surface whose elevation above the xy-plane is a function of x and y.

tessellation *See* TILING.

tesseract *See* HYPERCUBE.

tetartoid A polyhedron with 12 congruent faces, each an irregular pentagon. It has three different vertex figures.

tetracaidecadeltahedron A DELTAHEDRON with 14 faces and nine vertices.

tetracontagon A polygon with 40 sides.

tetractys The numbers 1, 2, 3, and 4, which sum to 10 and can be represented by a triangle of 10 dots. The tetractys was held to be sacred by the Pythagoreans.

tetracuspid *See* ASTROID.

tetragon *See* QUADRILATERAL.

tetragonal disphenoid A polyhedron with four congruent isosceles triangular faces and with congruent solid angles.

tetrahedral group The GROUP of DIRECT SYMMETRIES of a tetrahedron; it has 12 elements.

tetrahedral pentagonal dodecahedron *See* TETARTOID.

tetrahedron A polyhedron with four congruent faces, each of which is an equilateral triangle. Sometimes, any polyhedron with four faces.

tetrahemihexahedron A NONCONVEX UNIFORM heptahedron with four triangular faces and three square faces. It is a faceted octahedron.

tetrakaidecahedron A polyhedron with 14 faces.

tetrakis hexahedron A polyhedron with 24 triangular faces. It is the dual of the truncated octahedron.

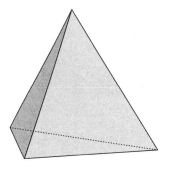

Tetrahedron

tetraktys *See* TETRACTYS.

tetriamond A polyiamond made up of four equilateral triangles.

tetromino A POLYOMINO made up of four squares.

theodolite An instrument used by surveyors to measure angles.

theorem A statement of a mathematical result in an axiomatic system that can be proven using the axioms of the system as well as other theorems that have already been proven.

Thiessen polygon *See* VORONOI REGION.

third curvature For a point on a curve, the value of $\sqrt{k^2 + \tau^2}$, where k is curvature and τ is torsion.

third proportional For numbers a and b, a number x such that $a/b = b/x$.

three-dimensional crystallographic group The GROUP of all SYMMETRY TRANSFORMATIONS of a given three-dimensional periodic pattern. There are 230 such groups.

tile *See* TILING.

tile-transitive tiling *See* ISOHEDRAL TILING.

tiling A covering of the plane or other surface by polygons, called tiles, so that every point on the surface is covered by a tile and two tiles intersect only along their edges. A vertex of the tiling is a point where three or more edges meet.

TIN *See* TRIANGULATED IRREGULAR NETWORK.

tip *See* TERMINAL POINT.

tomahawk A shape consisting of a semicircle and handle that can be used to trisect angles.

tombstone *See* HALMOS SYMBOL.

topological invariant Something, such as a number, topological property, or algebraic expression, left unchanged by a homeomorphism.

topological mapping *See* HOMEOMORPHISM.

topological quantum field theory The study of quantum field theories from physics in terms of the topology of spacetime.

topological space A set of points together with a specified collection of OPEN SETS that are closed under finite intersections and arbitrary unions and include the whole set and the empty set.

topologically edge transitive *See* HOMEOTOXAL TILING.

topologically equivalent *See* HOMEOMORPHIC.

topologically tile transitive *See* HOMEOHEDRAL TILING.

topologically vertex transitive *See* HOMEOGONAL TILING.

topology (1) The collection of OPEN SETS belonging to a topological space. (2) The study of properties of shapes and spaces that remain invariant under homeomorphisms. Topology is often called "rubber-sheet geometry" because it studies properties that do not change when shapes are stretched or shrunk.

toroidal Related to a TORUS. For example, a toroidal map is a map defined on the surface of a torus.

torque   Turning force. A torque vector is the CROSS PRODUCT of a force vector with the position vector of the point where the force is applied.

torsion   A measure of how a curve turns in space. The torsion of a planar curve is 0.

torus   A closed surface shaped like the surface of a doughnut. The plural is *tori*.

torus knot   A KNOT that can be drawn on the surface of a torus without self-intersections. *See also* (p, q)-TORUS KNOT.

torus link   A LINK that can be drawn on the surface of a torus without self-intersections.

total angular deficit   The sum of the angular deficits for all the vertices of a polyhedron.

total curvature   For an arc of a curve, the total curvature is the net change in the direction of a unit tangent vector from one endpoint to the other. For example, the total curvature of a semicircle traversed counterclockwise is 180° and the total curvature of a semicircle traversed clockwise is –180°. For a surface, the total curvature is the GAUSSIAN CURVATURE.

total space   *See* FIBER BUNDLE.

touch   To be tangent to.

trace   The sum of the elements along the main diagonal of a square matrix.

tracing puzzle   The problem of finding an EULER PATH for a given GRAPH, usually one that has an interesting or complex form.

track line   The actual path a ship or plane travels along.

tractrisoid   *See* PSEUDOSPHERE.

tractrix   The path traced out by a weight on a string being dragged by a person walking at a constant rate along the x-axis. It is an involute of the catenary. It can be given parametrically by $x = a(t - \tanh t)$, $y = a \operatorname{sech} t$, and it has the x-axis as an asymptote.

tractroid   The surface formed by rotating a TRACTRIX about its asymptote.

train   A series of consecutive circles lying between two nonintersecting circles, one of which contains the other. Each circle in the series is tangent to the two given circles and to one or two other circles in the series.

trajectory   The path of an object, usually given by parametric equations.

Torus

trammel of Archimedes The locus of a point fixed on a segment that slides with its endpoints fixed on two perpendicular lines.

transcendental number An irrational number that cannot be the root of a polynomial equation with rational coefficients. For example, π is a transcendental number.

transformation A function that has the properties of both a ONE-TO-ONE FUNCTION and an ONTO FUNCTION.

transformation, method of A method of solving a problem by solving instead the problem obtained by transforming the givens by an appropriate transformation.

transitive relation A relation, such as equality or less than, with the property that if a is related to b and b is related to c, then a is related to c.

transitivity class In a tiling, the set of all tiles or vertices equivalent to a specific tile or vertex of a tiling.

translation An ISOMETRY that moves every point in same direction by the same distance.

translation unit A region whose images under translation will cover a design.

translational symmetry Symmetry with respect to TRANSLATION.

transpose The matrix formed from a given square matrix A by interchanging the rows with the columns. It is denoted A^T.

transposition A permutation that interchanges two elements.

transvection *See* SHEAR.

transversal A line intersecting two other lines at distinct points.

transverse axis A segment connecting the two vertices of a hyperbola. *See also* CONJUGATE AXIS.

transverse cylindrical projection A CYLINDRICAL PROJECTION of a sphere onto a cylinder that is tangent along a LONGITUDE of the sphere.

transverse Mercator projection A projection of the earth's surface onto a cylinder whose axis is perpendicular to the polar axis and which is tangent to a MERIDIAN near the center of a particular region.

trapezium A quadrilateral with no parallel sides. In England, a quadrilateral with two parallel sides is called a trapezium.

trapezohedron A polyhedron whose faces are quadrilaterals.

trapezoid A quadrilateral in which two opposite sides are parallel.

Trapezoid Axiom The axiom that says that for three noncollinear points A, B, and C and any point B' on AB, the line through B' parallel to BC must intersect AC.

trapezoidal decahedron A CONVEX polyhedron with 10 faces. It is the dual of the pentagonal antiprism.

trapezoidal hexecontahedron A polyhedron with 60 kite faces. It is the dual of the RHOMBICOSIDODECAHEDRON.

trapezoidal icositetrahedron A polyhedron with 24 kite faces. It is the dual of the RHOMBICUBOCTAHEDRON.

trapezoidal octahedron A CONVEX polyhedron with eight kite faces. It is the dual of the square antiprism.

traveling salesperson path The shortest CLOSED PATH incident to every vertex in a WEIGHTED GRAPH.

trebly asymptotic triangle In HYPERBOLIC GEOMETRY, three lines, each of which is parallel to the other two. A trebly asymptotic triangle has no vertices.

tree A CONNECTED GRAPH that does not contain any circuits.

trefoil (1) A simple KNOT with three CROSSINGS; it exists in left-hand and right-hand forms. (2) A MULTIFOIL drawn outside of an EQUILATERAL TRIANGLE.

triacontagon A POLYGON with 30 SIDES.

triacontahedron A POLYHEDRON with 30 FACES.

triad A triple.

triadic von Koch curve *See* KOCH SNOWFLAKE.

triakisicosahedron A polyhedron with 60 triangular faces and 32 vertices. It is the dual of the truncated dodecahedron.

triakisoctahedron A NONCONVEX polyhedron with 24 triangular faces. It is the dual of the truncated cube.

triakistetrahedron A polyhedron with 12 triangular faces and eight vertices. It is the dual of the truncated tetrahedron.

triangle A polygon with three vertices and three edges.

Triangle Inequality The statement that for three points P, Q, and R, the distance from P to R is less than or equal to the distance from P to Q plus the distance from Q to R.

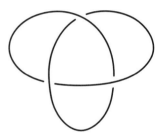

Right-hand trefoil

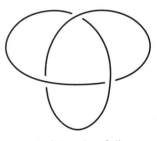

Left-hand trefoil

Trefoil knots

Triangle Law The addition law for vectors that says that the sum of vectors **v** and **w** is the third side of the triangle with sides **v** and **w**.

triangle of reference The triangle whose vertices are the three fixed points used to determine barycentric coordinates in a plane.

triangular dipyramid A convex polyhedron with six triangular faces and five vertices. It is a deltahedron and the dual of the triangular prism.

triangular matrix A square matrix all of whose elements either above or below the main diagonal are 0.

triangular orthobicupola A polyhedron with eight equilateral triangle faces and six square faces. It is space-filling.

triangular region All points in the interior of a triangle.

triangulate To find a triangulation of a surface or other region or shape.

triangulated irregular network (TIN) A partition, or triangulation, of a polygonal planar region into triangles that meet only at vertices and sides. A TIN is a two-dimensional cell complex.

triangulated manifold A manifold that has been triangulated.

triangulation The breaking up of a surface into triangles such that edges and vertices of adjacent triangles coincide. A triangulation of a higher-dimensional shape breaks it into simplices of the same dimension as the shape.

tridecagon A polygon with 13 sides.

trident The graph of $xy = ax^3 + bx^2 + cx + d$, $a \neq 0$, where a, b, c, and d are constant. This is also called the trident of Newton.

trigonal rotation A rotation through 120°.

trigonometric form The polar form of a complex number.

trigonometric series An infinite sum of sine and cosine functions.

trigonometry The study of right triangles and the relationships among the measurements of their angles and sides.

trihedral tiling A tiling in which each tile is congruent to one of three distinct prototiles.

trihedron A set of three concurrent, but not coplanar, lines (the edges of the trihedron) and the planes (the faces of the trihedron) that they determine. Sometimes, a trihedron is a solid angle consisting of three rays. *See also* MOVING TRIHEDRON.

trilinear coordinates Coordinates with respect to a grid of equilateral triangles.

trilinear polar For a given triangle with three concurrent cevians, construct the triangle whose vertices are the intersections of the cevians with the sides of the given triangle. Extend each side of this new triangle until it intersects the extended side of the given triangle that is opposite it. The line passing through these three points is the trilinear polar of the given triangle.

trilinear pole For a given trilinear polar, the trilinear pole is the point of concurrency of the three cevians of the triangle that determines it.

trimorphic tile A tile that admits exactly three different monohedral tilings.

triple scalar product The scalar $(\mathbf{u} \times \mathbf{v}) \cdot \mathbf{w}$, usually denoted $(\mathbf{u}\ \mathbf{v}\ \mathbf{w})$ for vectors $\mathbf{u}$, $\mathbf{v}$, and $\mathbf{w}$. It is 0 for three coplanar vectors.

triple vector product For vectors $\mathbf{u}$, $\mathbf{v}$, and $\mathbf{w}$, the triple vector product is the vector $(\mathbf{u} \times \mathbf{v}) \times \mathbf{w}$.

trirectangle A triangle on a sphere with three right angles or a quadrilateral in hyperbolic geometry with three right angles.

trirectangular *See* TRIRECTANGLE.

trirectangular tetrahedron A polyhedron with four faces, three of which are right triangles.

trisection of an angle The construction an angle with one-third the measure of a given angle. One of the three famous problems of antiquity, this problem is impossible to solve with compass and straightedge but can be solved using the Archimedes spiral, for example.

trisectrix A curve that can be used to trisect any angle.

triskaidecagon *See* TRIDECAGON.

triskelion A simple design with threefold cyclic symmetry.

tritangent circle The incircle or an excircle of a triangle; so called because they are tangent to all three sides of the triangle.

trivial As simple as possible and usually not of great interest. For example, the trivial symmetry group contains only the identity, and a trivial vector space is just a point.

trivial knot *See* UNKNOT.

trivial subspace A 0-dimensional subspace of a vector space. It contains only the zero vector.

trochoid The locus of a point rigidly attached to a curve that rolls along another curve. Examples are the cycloid, epicycloid, and hypocycloid.

true course A course that has been planned using true north rather than magnetic north.

truncate To cut off; for example, to cut off the corners of a polyhedron or the digits of a decimal expansion.

truncated cube A semiregular polyhedron with eight triangular faces and six octagonal faces.

truncated cuboctahedron *See* RHOMBITRUNCATED CUBOCTAHEDRON.

truncated dodecahedron A semiregular polyhedron with 20 triangular faces and 12 decagon faces.

truncated icosahedron A semiregular polyhedron with 12 pentagonal faces and 20 hexagonal faces.

truncated isosidodecahedron *See* GREAT RHOMBICUBOCTAHEDRON.

truncated octahedron A semiregular polyhedron with 14 faces: six square faces and eight regular hexagonal faces. It is a space-filling zonohedron.

truncated polyhedron A polyhedron formed by cutting off the vertices of another polyhedron.

truncated tetrahedron A semiregular polyhedron with four triangular faces and four hexagonal faces.

Tucker circle If a hexagon is inscribed in a triangle so that three alternate sides are parallel to the sides of the triangle and the other three alternate sides are ANTIPARALLEL to the sides of the triangle, the six vertices of the hexagon all lie on a circle called the Tucker circle.

twin cube A polyhedron with 36 faces formed from two congruent cubes having a space diagonal in common.

twist (1) An isometry of three-dimensional space that is the composite of a rotation and a translation along the axis of rotation. (2) *See* WRITHE NUMBER.

twisted curve A curve that does not lie in a plane.

twisted *q*-prism A solid formed from a prism with base a regular polygon with *q* vertices that has been sliced through the center parallel to the base with one half rotated through an angle of $(180/q)°$ and then reattached. Its symmetries include rotary reflection.

twisting A transformation of a space curve that preserves arc length and curvature but not torsion.

twisting number A measure of how twisted a link is. For an oriented link composed of two parallel twisted strands, the twisting number is the linking number minus the writhe number. It is not a link invariant but depends on the diagram of the link.

two-color symmetry A symmetry of a perfectly colored design that has two colors; it either preserves or interchanges the two colors.

two-dimensional primitive form In projective geometry, a plane or a bundle.

ultra-ideal point In hyperbolic geometry, a point added to each line in a pencil of ultraparallels. Thus, ultraparallels intersect at an ultra-ideal point.

ultraparallel In hyperbolic geometry, a line through a given point that does not intersect a given line and will not intersect the given line even if rotated a small amount. In ABSOLUTE GEOMETRY, a line that does not intersect a given line but is not parallel to it.

umbilic A point on a surface with the property that the normal curvature is the same in all directions.

undecagon *See* HENDECAGON.

undefined term A mathematical object in an axiomatic system that is not given a definition but whose properties are given by axioms. *See* AXIOMATIC SYSTEM.

undercrossing An ARC of a KNOT below a CROSSING in the KNOT DIAGRAM.

underdetermine To place so few conditions on the construction of an object that there are many different objects satisfying the condition.

underpass *See* UNDERCROSSING.

unfolding of a singularity The embedding of an unstable system in a stable system.

uniform network A network in which all of the solid angles are congruent.

uniform polyhedron A polyhedron in which each vertex is surrounded by the same arrangement of faces and edges.

uniform tiling A tiling in which each vertex is surrounded by the same arrangement of tiles.

uniformly bounded Describing a collection of sets with a nonzero INPARAMETER and a finite CIRCUMPARAMETER.

unilateral tiling A tiling in which each edge of the tiling is a complete side of at most one tile.

union The set containing all elements in either of two sets. The notation $S \cup T$ means the union of the set S with the set T.

unit binormal A binormal vector of length 1. *See* BINORMAL VECTOR.

unit cell *See* LATTICE UNIT.

unit circle A circle with radius equal to 1, usually centered at the origin of the Cartesian plane.

unit cube A cube with edges of length 1.

unit lattice A plane lattice that is generated by a parallelogram with area equal to 1 or a space lattice that is generated by a parallelepiped with volume equal to 1.

unit normal vector A normal vector with length 1. *See* NORMAL VECTOR.

unit point The point in the projective plane with homogeneous coordinates $(1, 1, 1)$.

unit square A square with sides of length 1.

unit vector A vector with magnitude 1.

unitary matrix A square matrix with complex elements such that A^T is the complex conjugate of A^{-1}.

universal cover A set in a Euclidean space that can contain every set in that space having diameter no greater than 1.

universal covering space A simply connected COVERING SPACE; so called because it covers any other covering space of its base space.

unknot A KNOT that can be DEFORMED into a CIRCLE.

unknotting number For a given KNOT DIAGRAM, the unknotting number is the least possible number of CROSSING switches, either from left-hand to right-hand or vice versa, that change the knot diagram to the diagram of the UNKNOT.

unlink A LINK that can be deformed into a collection of disjoint circles.

unstable Dramatically changed by small changes in parameters or input.

valence The number of edges incident to a vertex in a tiling or graph. If every vertex in a tiling or graph has valence n, the tiling or graph is said to be n-valent.

van Hiele model A model proposed by Pierre M. van Hiele and Dina van Hiele-Geldorf used in mathematics education that gives the levels of thinking that students pass through as they learn geometry. The stages of the van Hiele model are visual, analysis, informal deduction, formal deduction, and rigor.

vanishing point A point on the picture plane of a perspective projection where lines parallel in the three-dimensional world appear to meet on the picture plane. Each class of parallel lines determines one vanishing point, which is the intersection of the image plane with the line in the class that passes through the center of projection. A vanishing point lies on the horizon line of the picture.

Varignon parallelogram The parallelogram whose sides are the medians of a quadrilateral.

Veblen-Young axiom The Veblen-Young axiom states that if a line intersects two sides of a triangle, then it will intersect the third side when it is extended. Parallel lines do not exist in a geometry that satisfies this axiom.

vector An object having magnitude and direction. The magnitude is given by a real number and the direction is given relative to a fixed axis. A vector starting at the origin in the Cartesian plane is usually given by the coordinates of its endpoint. Two vectors are equal if they have the same magnitude and direction, even if they have different locations. More generally, a vector is an element of a vector space. A vector is denoted by a boldface letter or by a letter with an arrow over it.

vector difference The vector obtained by subtracting one vector from the other.

vector field An assignment of a vector to each point in a region.

vector polynomial A polynomial whose coefficients are vectors instead of numbers.

vector product *See* CROSS PRODUCT.

vector space A set of elements, called vectors, that has two operations, vector addition and scalar multiplication. In vector addition, two vectors are added together to give a vector. A vector space is an

ABELIAN GROUP under vector addition. In scalar multiplication, a vector is multiplied by a scalar to give a vector. Scalar multiplication satisfies the following properties of associativity and distributivity, for scalars s and t and vectors $\mathbf{v}$ and $\mathbf{w}$: $s(t\mathbf{v}) = (st)\mathbf{v}$; $(s + t)\mathbf{v} = s\mathbf{v} + t\mathbf{v}$; and $s(\mathbf{v} + \mathbf{w}) = s\mathbf{v} + s\mathbf{w}$.

vector sum A vector that is the sum of two or more vectors. It can be obtained by taking the tail of one of the vectors and placing it at the tip of the other.

velocity Change in position with respect to time; velocity is given as a real number in units of distance per time.

velocity field A vector field whose vectors are velocity vectors.

velocity vector A vector giving the direction and magnitude of change in position with respect to time.

vers *See* VERSED SINE.

versed cosine For angle A, the versed cosine is $1 - \sin A$.

versed sine For angle A, the versed sine is $1 - \cos A$.

versiera *See* WITCH OF AGNESI.

versine *See* VERSED SINE.

vertex A point, usually the endpoint of a segment or arc, contained in a geometric figure.

vertex coloring A coloring of each vertex of a graph.

vertex figure A vertex figure at a vertex of a polygon is the segment joining the midpoints of the sides adjacent to the vertex. A vertex figure at a vertex of a polyhedron or tiling is the polygon whose sides are segments connecting the midpoints of each pair of adjacent edges that are incident with the vertex.

vertex neighborhood The region consisting of all tiles incident to a given vertex of a tiling of the plane.

vertex of a pencil The point of concurrency of the lines in a pencil.

vertex-regular Describing a polygon, polyhedron, or tiling, whose vertices all have regular vertex polygons.

vertical angles Two nonadjacent angles formed at the intersection of two lines. They are equal.

Vertex figures at vertex *A* for a hexagon, a hexagonal tiling, and a cube

vertical line test A test used to determine whether or not a curve in the Cartesian plane is the graph of a function. If every vertical line cuts the curve in at most one point, the curve is the graph of a function. Otherwise, it is not.

vesica piscis The convex region interior to two congruent circles such that the center of each circle is on the circumference of the other.

viewpoint A fixed point on a TERRAIN.

visibility function A function that assigns 1 to each pair of points in a TERRAIN if they are mutually visible and 0 otherwise (the point-point visibility function). Also, a function that assigns 1 to a point and a region if the region is visible from the point and 0 otherwise (the point-region visibility function).

visibility graph A graph whose vertices correspond to points on a TERRAIN and whose edges connect mutually visible points.

visibility index The number of vertices in a VISIBILITY GRAPH that are visible from a given viewpoint.

visibility map For a given viewpoint, a projection of a TERRAIN onto the xy-plane showing the boundaries of the visible and invisible regions of the terrain.

visibility matrix The INCIDENCE MATRIX of a VISIBILITY GRAPH.

visibility region *See* VISIBLE REGION.

visibility-equivalent Describing TERRAINS having the same VISIBILITY MAPS.

visible point A visible point is a point in a TERRAIN that can be viewed from a viewpoint on the terrain. Thus, a segment from the viewpoint to a visible point lies above the terrain.

visible region That part of a TERRAIN that can be viewed from a given viewpoint. Thus, a segment from the viewpoint to a point in the visible region lies above the terrain.

volume A measure of the size of a region of three-dimensional space.

volute A spiral used at the top of the Ionic column in Greek architecture. *See also* LITUUS.

Voronoi diagram For a given set of points, called sites or seeds, a subdivision of the plane into convex regions, called Voronoi regions, such that each Voronoi region contains exactly one of the sites and every point in the interior of a Voronoi region is closer to its site than to any other site.

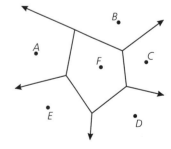

Voronoi diagram for points *A*, *B*, *C*, *D*, *E*, and *F*

Voronoi edge An edge of a Voronoi region contained in a VORONOI DIAGRAM. A point on a Voronoi edge has two nearest sites.

Voronoi polygon *See* VORONOI REGION.

Voronoi region *See* VORONOI DIAGRAM.

Voronoi vertex A vertex of a VORONOI REGION. A Voronoi vertex has three or more nearest sites.

vortex flow A flow that goes in circles around a VORTEX POINT.

vortex point An equilibrium point of a dynamical system that is at the center of nested cycles of the system.

waist A moveable joint on the support of the shoulder of a ROBOT ARM.

wallpaper group The GROUP of all SYMMETRY TRANSFORMATIONS of a given tiling. There are 17 possible wallpaper groups.

wallpaper pattern *See* TILING.

Wang tiles Square tiles with colored edges. To form a tiling with these tiles, edge colors must match on adjacent tiles and tiles may not be rotated or reflected.

watershed The boundary between two different BASINS.

Watt mechanism A simple mechanism with four rods used to trace out a curve that closely approximates a straight segment.

wavelet A function that can be combined together with transformations of itself to give more complicated functions.

web diagram A diagram, which looks like a spiderweb, used to study the iterations of a function $y = f(x)$. The segments of the web connect points $(x, f(x))$ to $(f(x), f(x))$ or $(f(x), f(x))$ to $(f(x), f(f(x)))$, and so on.

weight In a GRAPH, a number assigned to an edge. For example, the weight could represent the length of the edge or the construction cost of a road represented by the edge.

weight matrix An $n \times n$ matrix that represents a WEIGHTED GRAPH with n vertices and at most one edge connecting each pair of vertices. The entry in the ith row and jth column is the weight of the edge connecting the ith vertex to the jth vertex; it is ∞ if there is no edge connecting the ith vertex to the jth vertex; and it is 0 if $i = j$.

weighted adjacency matrix *See* WEIGHT MATRIX.

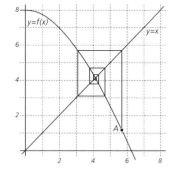

Web diagram

weighted graph A GRAPH in which each edge has a real number, or weight, assigned to it.

well-oiled compass *See* COLLAPSING COMPASS.

wheelbarrow A POLYIAMOND made up of 18 triangles that is shaped like a wheelbarrow. It admits one MONOHEDRAL TILING of the plane but does not admit any ISOHEDRAL TILINGS.

Whitehead link A LINK with two components that cannot be separated even though the linking number is 0.

width The width of a rectangle is the length of its base. In general, the width of closed curve, region, or surface is the distance between parallel tangent lines or planes. A shape can have different widths in different directions.

Wigner-Seitz region *See* VORONOI REGION.

wild knot A KNOT whose KNOT DIAGRAM has infinitely many CROSSINGS.

wind vector A vector giving the direction and speed of the wind.

wind-drift angle The angle between the heading of a ship or plane and the direction of the wind.

wind-drift triangle A vector triangle whose sides represent the velocity of the wind, the velocity of a boat or airplane relative to the air and the resultant, which represents the actual velocity of the object with respect to the ground.

winding number For a closed curve, the number of revolutions the curve makes about a given fixed point when the curve is traversed once in the counterclockwise direction.

witch of Agnesi The curve given by the equation $x^2 y = 4a^2(2a - y)$, with a constant.

workspace *See* REACHABILITY REGION.

world coordinates Coordinates with respect to a WORLD FRAME.

world frame A fixed frame.

world line In physics, the curve describing the trajectory of a particle in SPACETIME GEOMETRY.

world point A point in a three-dimensional space that is to be projected onto a two-dimensional surface.

world surface In STRING THEORY, the surface swept out by a string as it moves through spacetime.

wrench A pair consisting of a force vector and a TORQUE vector.

writhe number A measure of how twisted a KNOT is. For an ORIENTED KNOT, the writhe number is the sum of the signs of the CROSSINGS in the KNOT DIAGRAM. It is not a KNOT INVARIANT.

x-axis The horizontal axis in the Cartesian plane.

x-intercept The point of intersection of a line or curve with the x-axis.

y-axis The vertical axis in the Cartesian plane.

y-intercept The point of intersection of a line or curve with the y-axis.

Young's geometry A FINITE PROJECTIVE GEOMETRY consisting of nine points and 12 lines.

z-axis The axis perpendicular to the xy-plane in a three-dimensional coordinate system; it is usually drawn vertically.

Zeeman catastrophe machine A simple device made from cardboard discs and rubber bands that gives a physical demonstration of a CATASTROPHE.

zero angle An angle equal to $0°$.

zero matrix A matrix all of whose elements are 0. It is denoted **0**.

zero vector A vector having length equal to 0. It is denoted **0**.

zone (1) The collection of all faces of a ZONOHEDRON with edges parallel to a given edge. (2) The zone of a line that is not contained in an ARRANGEMENT OF LINES is the set of cells in the arrangement that intersect the line. *See* ARRANGEMENT.

zonohedron A zonohedron has faces that are polygons with an even number of sides and whose opposite sides are parallel. In some contexts, a zonohedron is a convex polyhedron whose faces are parallelograms.

SECTION TWO
BIOGRAPHIES

Abbott, Edwin Abbott (1838–1926) British clergyman who wrote *Flatland*, a satire of British society that demonstrated the nature of HIGHER-DIMENSIONAL geometry in an appealing and comprehensible way.

Abel, Niels Henrick (1802–29) Norwegian mathematician who contributed to all areas of mathematics. He proved that the general quintic polynomial equation could not be solved using radicals and studied commutative groups, now called Abelian groups in his honor.

Abū al-Wafā al Būjāni (940–998) Persian mathematician and astronomer in the caliph's court in Baghdad whose contributions to TRIGONOMETRY were of great use to astronomers. He systematized the study of the six trigonometric functions, made accurate tables of their values, and proved new theorems in SPHERICAL TRIGONOMETRY, including the law of sines for spherical triangles. He discovered the REGULAR and QUASIREGULAR tessellations of the sphere.

Agnesi, Maria Gaetana (1718–99) Italian mathematician who is noted for her study of the versiera curve, now called the WITCH OF AGNESI. A child prodigy, Agnesi mastered many languages, including Latin and Greek, by the age of nine. To help her brother learn CALCULUS, she wrote a book that later become very influential in the development of calculus in Italy.

Alberti, Leon Battista (1404–72) Italian mathematician and writer of the Renaissance who was one of the first to master the art of PERSPECTIVE. He introduced the concept of the cone of vision and wrote the earliest book on perspective drawing, *Della pittura.*

Aleksandrov, Pavel (1896–1982) Russian mathematician who was a major figure in the early development of POINT SET TOPOLOGY and ALGEBRAIC TOPOLOGY. He studied at Moscow University and later became a professor there.

Alembert, Jean Le Rond d' (1717–83) French mathematician who was the first to solve the DIFFERENTIAL EQUATION describing a vibrating string. The orphaned son of a French military officer, d'Alembert was brought up by a glassblower. He had wide interests and talents and was a founder of the theory of

Maria Agnesi (The Granger Collection, New York)

functions of a complex variable and the theory of PARTIAL DIFFERENTIAL EQUATIONS.

Alexander, James (1888–1971) American mathematician and professor at Princeton University who was a leading researcher in topology, knot theory, and ALGEBRAIC TOPOLOGY. He discovered the ALEXANDER HORNED SPHERE, the ALEXANDER POLYNOMIAL, and the Alexander duality theorem.

Amman, Robert (1948–92) American amateur mathematician who made many discoveries in the study of PERIODIC and APERIODIC TILINGS that bear his name.

Apastamba (c. 600 B.C.E.) Vedic seer who wrote one of the main Sulba Sutras, which are the oldest known writings on geometry. His work includes the solution of a general linear equation, an approximation for $\sqrt{2}$, and an approximate squaring of the circle.

Apollonius of Perga (c. 240–c. 174 B.C.E.) Greek mathematician and philosopher who wrote a comprehensive treatise on the CONIC SECTIONS, the *Conics,* which became the standard source of properties of the conic sections and was used by KEPLER, DESCARTES, and NEWTON.

Appel, Kenneth (b. 1932) American mathematician who solved the FOUR-COLOR PROBLEM together with his colleague WOLFGANG HAKEN at the University of Illinois. Their proof required more than 1,000 hours of computer calculations.

Archimedes (c. 287–212 B.C.E.) Greek physicist and engineer; the greatest mathematician of ancient times. He found the surface area and volume of the sphere, gave an accurate estimate for the value of pi, and used a spiral to solve the angle trisection problem. Archimedes also studied optics, invented many practical devices, founded the science of hydrostatics, and formulated several axioms that enriched and refined Euclidean geometry.

Archytas (c. 438–c. 347 B.C.E.) Greek Pythagorean who solved the Delian problem of duplicating the cube by using the intersection of three solids. He was a teacher of both PLATO and EUDOXUS.

Archimedes (Alinari/Art Resource, NY)

Argand, Jean Robert (1768–1822) French mathematician of Swiss origin who developed the geometric interpretation of the COMPLEX NUMBERS that is still in use today.

Aristaeus the Elder (c. 330 B.C.E.) Greek mathematician who was one of the first to study the CONIC SECTIONS.

Aristotle (384–322 B.C.E.) Greek philosopher who studied the mathematical concepts of INCOMMENSURABILITY, continuity, and the infinite. Aristotle's recognition of the importance of unproved statements in the form of COMMON NOTIONS and POSTULATES as the starting point of a mathematical proof has had a lasting influence on mathematicians.

Arnol'd, Vladimir Igorevich (b. 1937) Russian mathematician who has made major contributions to the study of DYNAMICAL SYSTEMS. Arnol'd now holds a joint appointment at the Steklov Mathematical Institute in Russia and Université Paris 9 in France.

Artin, Emil (1898–1962) Austrian-born mathematician who studied in Germany and later taught in the United States and Germany. Known primarily as an algebraist, Artin created the theory of BRAIDS and braid groups, now widely used in KNOT theory.

Aryabhata, Kusumapura (476–c. 550) Indian astronomer who introduced the SINE and VERSINE functions to Indian trigonometry. He gave 3.1416 as an approximation of pi and made important contributions to algebra.

Atiyah, Michael (b. 1929) British mathematician of Lebanese descent who has made major contributions to ANALYSIS, GEOMETRY, TOPOLOGY, KNOT THEORY, DIFFERENTIAL EQUATIONS, and theoretical physics. Jointly with ISADORE SINGER, he proved the Atiyah-Singer index theorem, which connects the analytic and topological properties of MANIFOLDS. He was awarded the Fields Medal in 1966 and is now director of the Newton Institute in Cambridge, England.

Banchoff, Thomas (b. 1938) American mathematician who has pioneered the use of computer graphics to understand HIGHER-DIMENSIONAL geometry. He has introduced many artists, including Salvador Dali, to the beauty of geometry. Banchoff

Thomas Banchoff (Brown University)

received his doctorate from the University of California at Berkeley in 1964 and is now professor of mathematics at Brown University.

Barlow, William (1845–1934) English crystallographer who independently gave a list of the 230 CRYSTALLOGRAPHIC GROUPS in 1894.

Baudhayana (c. 800 B.C.E.) Vedic seer who wrote one of the main Sulba Sutras, the oldest known writings on geometry. His work includes approximate values for pi and √2.

Beltrami, Eugenio (1835–1900) Italian mathematician who worked in DIFFERENTIAL GEOMETRY and theoretical physics. He convinced other mathematicians of the validity of NON-EUCLIDEAN GEOMETRY by using the PSEUDOSPHERE, also called the Beltrami sphere, as a concrete model for HYPERBOLIC GEOMETRY.

Bernoulli, Daniel (1700–82) Swiss mathematician who contributed to many areas, including hydrodynamics, vibrating systems, probability, and fluid flow. The son of JOHANN BERNOULLI, Daniel spent six years working with LEONHARD EULER in Saint Petersburg, Russia.

Bernoulli, Jacob (1654–1705) Swiss mathematician, physicist, and astronomer who was an important contributor to the development of CALCULUS. He introduced POLAR COORDINATES and the LEMNISCATE OF BERNOULLI.

Bernoulli, Johann (1667–1748) Swiss mathematician and physicist who was the younger brother of JACOB BERNOULLI and an important contributor to the development of CALCULUS, probability, and the CALCULUS OF VARIATIONS and was the first to study the natural exponential function e^x. He was professor of mathematics at Groningen in Holland and later at the University of Basel in Switzerland.

Bernoulli, Johann, II (1710–90) Swiss mathematician who worked on the mathematical properties of light and heat. He was the son of JOHANN BERNOULLI.

Besicovitch, Abram Samoilovitch (1891–1970) Russian-born mathematician who is noted for his solution of the Kakeya

Johann Bernoulli (Image Select/Art Resource, NY)

problem, finding the smallest area needed to rotate a line segment through 360°. He also worked on almost periodic functions, geometric measure theory, and REAL and COMPLEX ANALYSIS. Shortly after the Russian Revolution, Besicovitch immigrated to England where he became professor of mathematics at Cambridge University.

Betti, Enrico (1823–92) Italian mathematician who worked in geometry, algebra, and TOPOLOGY. He developed the BETTI NUMBERS, which extend the Euler number to higher-dimensional shapes. Betti also worked for the unification of Italy and later served as a member of its parliament.

Bezier, Pierre (1910–99) French mechanical engineer who developed the BEZIER SPLINE for constructing a smooth curve from a few points of data. He was the director of the Machine Tool Division at Renault.

Bhāskara I (c. 600) Indian astronomer and mathematician who developed a formula for the SINE of an acute angle.

Bhāskara II (1115–c. 1185) Indian astronomer and mathematician who was noted for his work in algebra. He also wrote about SPHERICAL GEOMETRY and SPHERICAL TRIGONOMETRY and their applications to geography and astronomy and developed original concepts that were later used in CALCULUS.

Bing, R. H. (1914–86) American topologist and native of Texas who studied TOPOLOGY and two- and three-dimensional MANIFOLDS. The BING LINK is named after him.

Birkhoff, George David (1884–1944) American mathematician who did important work in ergodic theory, DYNAMICAL SYSTEMS, and quantum theory. By solving a three-body problem posed by Poincaré, he became the first American mathematician to have international fame.

Birman, Joan (b. 1927) American mathematician who is a leading expert in the theory of KNOTS and BRAIDS. She began her study of mathematics after raising three children and is now a professor at Barnard College.

Joan Birman (Joe Pineiro)

al-Birūni, Abū al-Rayhan (973–1048) Islamic mathematician, astronomer, and astrologer of central Asia who was also a skilled diplomat and traveled widely. He wrote about many subjects, including TRIGONOMETRY and SPHERICAL TRIGONOMETRY.

Bolyai, János (1802–60) Hungarian mathematician who independently discovered HYPERBOLIC GEOMETRY. He also developed ABSOLUTE GEOMETRY, which is based only on Euclid's first four postulates.

Bombieri, Enrico (b. 1940) Italian mathematician who studies NUMBER THEORY and MINIMAL SURFACES. He was awarded the Fields Medal in 1974 for work in number theory and is currently interested in applying number theory to the study of QUASICRYSTALS. He is a member of the Institute for Advanced Study in Princeton, New Jersey.

Bonnet, Pierre Ossian (1819–92) French engineer who later chose to become a teacher of mathematics. He worked in DIFFERENTIAL GEOMETRY and introduced the concept of GEODESIC CURVATURE, proving the general case of the Gauss-Bonnet theorem.

Richard Borcherds
(UC Berkeley, 1998)

Borcherds, Richard E. (b. 1959) British mathematician who was awarded the Fields Medal in 1998 for a dazzling theorem that connects symmetries in higher-dimensional space to modular functions and STRING THEORY. He now teaches at the University of California at Berkeley.

Bordoni, Antonio Maria (1789–1860) Italian mathematician who was interested primarily in DIFFERENTIAL GEOMETRY. He was a professor of mathematics at Pavia University.

Borel, Armand (1923–2003) Swiss-born American mathematician who made contributions to NUMBER THEORY, TRANSFORMATION GROUPS, and ALGEBRAIC TOPOLOGY. He was a member of the NICHOLAS BOURBAKI group and was professor emeritus at the Institute for Advanced Study in Princeton.

Bott, Raoul (b. 1923) Hungarian-born American mathematician who won numerous prizes for his work in TOPOLOGY and ALGEBRAIC TOPOLOGY. His most important contribution is

Armand Borel (IAS)

Raoul Bott (Kris Snibbe/ Harvard News Office)

the Bott periodicity theorem about the HOMOTOPY GROUPS of higher-dimensional spaces. He was a professor of mathematics at Harvard University for 40 years before retiring.

"Bourbaki, Nicolas" A group of French mathematicians started in 1935 who adopted the pseudonym "Nicholas Bourbaki" to publish their textbooks, which were to cover all of mathematics. Nicholas Bourbaki emphasized the importance of the abstract axiomatic approach to the study of mathematics. The Bourbaki approach influenced the theoretical development of mathematics during the 20th century.

Bouvelles, Charles de (c. 1471–c. 1553) French philosopher and geometer who introduced the HYPOTROCHOID as a way to solve the problem of SQUARING THE CIRCLE.

Boys, Charles Vernon (1855–1944) British physicist and inventor who studied the mathematical properties of soap bubbles.

Brahmagupta (598–c. 665) Indian astronomer and mathematician who wrote about arithmetic, algebra, TRIGONOMETRY, and mensurational geometry. He developed methods for approximating the sine function and discovered a formula for the area of a quadrilateral inscribed in a circle.

Branner, Bodil (b. 1943) Danish mathematician who has studied DYNAMICAL SYSTEMS defined by cubic polynomials and their associated FRACTALS. She is a professor of mathematics at the Technical University in Copenhagen.

Bravais, Auguste (1811–63) French scientist who independently discovered the 32 CRYSTAL CLASSES and the 14 three-dimensional LATTICES that are named for him.

Brianchon, Charles Julien (c. 1783–1864) French mathematician who served in Napoleon's army before beginning a teaching career in mathematics. While a student at the École Polytechnique, Brianchon proved Brianchon's theorem, the dual of Pascal's MYSTIC HEXAGRAM THEOREM. In 1820, Brianchon and Poncelet gave the first proof of the properties of the NINE-POINT CIRCLE.

Brocard, Henri (1845–1922) French army officer and meteorologist who discovered the BROCARD POINTS and the BROCARD ANGLE. He also studied ALGEBRAIC CURVES.

Brouwer, L. E. J. (1881–1966) Dutch topologist and logician who was interested in the foundations of mathematics. He is famous for the Brouwer fixed point theorem, a cornerstone of modern topology.

Bruijn, N. G. de (b. 1918) Dutch mathematician who has made important contributions to the study of PERIODIC and APERIODIC TILINGS, geometry, NUMBER THEORY, combinatorics, computer science, mathematical language, and applied mathematics. He has been professor emeritus of the Technological University of Eindhoven since 1984.

Brunelleschi, Filippo (1377–1446) Italian architect and goldsmith who was the first Renaissance artist to use linear PERSPECTIVE. He invented the checkerboard grid as a tool for perspective.

Buffon, Georges (1707–88) French mathematician who studied probability and developed a probabilistic method for computing pi by throwing needles onto a tiled floor and counting how many of them landed on the lines of the tiling.

Calabi, Eugenio (b. 1923) Italian-born American mathematician who studies the DIFFERENTIAL GEOMETRY of COMPLEX MANIFOLDS. He is a member of the National Academy of Sciences and is professor of mathematics at the University of Pennsylvania.

Cantor, Georg (1845–1918) German mathematician who founded SET THEORY in order to solve some difficult problems in CALCULUS. Set theory is now used as the foundational theory for all of mathematics. He discovered the CANTOR SET, one of the earliest examples of a FRACTAL.

Carathéodory, Constantin (1873–1950) German mathematician of Greek heritage whose mathematical interests were in the CALCULUS OF VARIATIONS and its applications to geometric optics, PARTIAL DIFFERENTIAL EQUATIONS, and theoretical physics. Carathéodory studied in Germany and then taught in Turkey and Greece before returning to Germany.

Cartan, Élie (1869–1951) French mathematician who made major contributions to LIE GROUPS, Lie algebras, geometry, DIFFERENTIAL GEOMETRY, and TOPOLOGY. He collaborated with his son HENRI CARTAN, also a mathematician.

Cartan, Henri (b. 1904) French mathematician who worked in many areas of mathematics including analytic functions, HOMOLOGICAL ALGEBRA, and ALGEBRAIC TOPOLOGY. He is the son of the mathematician ÉLIE CARTAN and has been retired since 1975.

Cassini, Giovanni Domenico (1625–1712) Italian astronomer and hydraulic engineer who studied the curves known today as the OVALS OF CASSINI. He was born in Italy, where he studied astronomy and hydraulics. Invited to France in 1669, he became the director of the Paris Observatory in 1671.

Catalan, Eugène (1814–94) Belgian mathematician who was the first to study the DUALS of the SEMIREGULAR SOLIDS. A student and later professor at the École Polytechnique, he studied continued fractions and NUMBER THEORY and discovered the Catalan numbers while considering dissections of polygons into triangles by nonintersecting diagonals.

Cauchy, Augustin-Louis (1789–1857) French mathematician who made significant contributions to many areas of mathematics and physics, many of which bear his name. He furthered the development of CALCULUS and was a founder of the theory of complex functions.

Cavalieri, Francesco Bonaventura (1598–1647) An Italian Jesuit priest, Cavalieri discovered ways of finding areas and volumes of geometric figures that provided a foundation for the later development of calculus. A friend of Galileo, he was also interested in astronomy.

Cayley, Arthur (1821–95) British algebraist who contributed many tools of great importance to geometry, including coordinates and metrics in PROJECTIVE GEOMETRY and matrices and MATRIX multiplication. Cayley was a founder of LINEAR ALGEBRA and the study of higher-dimensional geometries and the first to write about the OCTONIANS, which are also called the Cayley numbers.

Cech, Eduard (1893–1960) Czech mathematician who contributed to DIFFERENTIAL GEOMETRY, COMBINATORIAL TOPOLOGY, and HOMOLOGY THEORY.

Ceva, Giovanni (1647–1734) Italian mathematician who also worked in economics, mechanics, and hydraulics. Ceva was professor of mathematics at the University of Mantua in Italy. His most important work was in geometry with the discovery of CEVA'S THEOREM.

Chasles, Michel (1793–1880) French geometer who made a fortune in banking before devoting his life to mathematics. He studied the CROSS RATIO and made significant contributions to ANALYTIC GEOMETRY and was professor of mathematics at the École Polytechnique in Paris.

Chern, Shiing-Shen (b. 1911) Chinese-born mathematician who studied in China, Germany, and France before coming to the United States. He has made many important contributions to DIFFERENTIAL TOPOLOGY, DIFFERENTIAL GEOMETRY, and theoretical physics. He is now professor emeritus at the University of California at Berkeley.

Clairaut, Alexis Claude (1713–65) French mathematician who studied CALCULUS, the CALCULUS OF VARIATIONS, DIFFERENTIAL EQUATIONS, and GEODESICS on QUADRICS. At age 18, he was the youngest person ever elected to the Paris Academy of Sciences. Using calculus and Newton's law of gravitation, he was able to compute the Moon's orbit and predicted the 1759 return of Halley's comet.

Shiing-Shen Chern
(UC Berkeley)

Clifford, William Kingdon (1845–79) English mathematician who studied NON-EUCLIDEAN GEOMETRY, CURVATURE of space, and Hamilton's QUATERNIONS. He is noted for his work on the non-Euclidean spaces known as Clifford-Klein spaces and the topology of spaces of CONSTANT CURVATURE.

Connelly, Robert (b. 1942) American geometer who is a leading researcher in the study of the rigidity of FRAMEWORKS. He is a professor at Cornell University and in 1977 discovered a closed, simply connected polyhedral surface that is not rigid.

Connes, Alain (b. 1947) French mathematician who founded the study of NONCOMMUTATIVE GEOMETRY. Connes has a joint appointment at Ohio State University and at the Institute des Hautes Études Scientifiques in France. He was awarded the Fields Medal in 1983 for work on HILBERT SPACES and operator algebras and the Crafoord Prize by the Royal Swedish Academy of Sciences in 2001 for his work on the theory of operator algebras and for founding noncommutative geometry.

Conway, John (b. 1937) British mathematician now living in the United States who has contributed to many areas of mathematics, including abstract algebra, NUMBER THEORY, game theory, combinatorics, KNOT THEORY, and geometry. A creative mathematician with penetrating insights, Conway is the inventor of surreal numbers, TANGLES, and the game of Life for computers. There are many mathematical concepts named for him, including the Conway group, the CONWAY CRITERION, and the CONWAY WORM. He is a professor of mathematics at Princeton University.

John Conway
(Robert Matthews)

Coxeter, H. S. M. (1907–2003) Canadian mathematician who was a great leader and teacher of geometry. He was born and educated in England and became a professor of mathematics at the University of Toronto in 1936, where he was professor emeritus. Coxeter made major contributions to the study of polyhedra in higher dimensions, NON-EUCLIDEAN GEOMETRY, GROUP THEORY, and combinatorics.

Cramer, Gabriel (1704–52) Swiss mathematician who is known today for Cramer's rule for solving a system of linear equations. Cramer received his doctorate at the age of 20, then accepted a teaching position in Geneva that allowed him to travel widely and continue his study of mathematics.

Dandelin, Germinal (1794–1847) French-born Belgian mathematician and engineer who discovered the DANDELIN SPHERES that are tangent to the plane determining a conic at the focus of the conic.

Dantzig, George (b. 1914) American mathematician who invented the simplex algorithm for solving LINEAR PROGRAMMING problems. He is professor emeritus at Stanford University.

Darboux, Gaston (1842–1917) French mathematician who made important contributions to DIFFERENTIAL GEOMETRY and ANALYSIS. He studied the CYCLIDES and the shortest path between two points on a surface.

Daubechies, Ingrid (b. 1954) American mathematician who is a founder of the theory of WAVELETS. Born in Belgium, Daubechies was awarded a doctorate in physics and received a MacArthur fellowship in 1992. She is a member of the National Academy of Sciences and a professor at Princeton University.

Dedekind, Richard (1831–1916) German analyst who developed the DEDEKIND CUT, a precise way of constructing the REAL NUMBERS from the RATIONAL numbers using the language of SET THEORY.

Dehn, Max (1878–1952) German mathematician who developed the procedure known as DEHN SURGERY in topology. When he was a student of DAVID HILBERT at Göttingen, he solved the third of the problems proposed by Hilbert in 1900. In 1914, he proved that the left-hand and the right-hand TREFOIL knots are not equivalent to each other. In 1940, he came to the United States, where he taught at Black Mountain College in North Carolina.

Delaunay (Delone), Boris Nikolaevich (1890–1980) Russian mathematician who studied algebra, NUMBER THEORY, and the GEOMETRY OF NUMBERS. He introduced the DELAUNAY TRIANGULATION in the context of his work on the structure of crystals.

Democritus (c.460–c. 379 B.C.E.) Ancient Greek philosopher who discovered formulas for the volume of a prism, pyramid, cylinder, and cone.

Desargues, Girard (1591–1661) French architect and military engineer who made many significant discoveries in PROJECTIVE GEOMETRY, including the theorem named after him. The value of his work was recognized only many years after his death.

Descartes, René (1596–1650) French philosopher, scientist, and mathematician who was interested in the deepest and most universal principles of knowledge and how they could be

Ingrid Daubechies
(Denise Applewhite)

René Descartes (Réunion des Musées Nationaux/ Art Resource, NY)

discovered. His invention of COORDINATE GEOMETRY and ANALYTIC GEOMETRY allowed the powerful techniques of symbolic algebra to be used to uncover and understand properties of geometric shapes. Today, analytic geometry is an essential tool for mathematicians and the coordinate plane is called the Cartesian plane to honor Descartes for his discovery

Devaney, Robert (b. 1948) American mathematician who has made significant contributions to the study of FRACTALS and DYNAMICAL SYSTEMS. He is a professor of mathematics at Boston University, where he has developed computer graphics software for viewing and studying fractals.

Diocles (c. 210–190 B.C.E.) Greek mathematician who studied the CONIC SECTIONS. He discovered the properties of the CISSOID and discovered the relationship between the focus and the directrix of a PARABOLA.

Dirichlet, Peter Lejeune (1805–59) Of Belgian ancestry, Dirichlet was raised in Germany, studied in France, and returned to Germany to teach at the University of Berlin and later at Göttingen. He worked in NUMBER THEORY, DIFFERENTIAL EQUATIONS, and FOURIER SERIES. He initiated the development of analytic number theory and proposed the definition of FUNCTION that is used today. He originated the pigeonhole principle, which states that if n objects are placed in fewer than n pigeonholes, at least one pigeonhole must contain more than one object.

Donaldson, Simon K. (b. 1957) British mathematician who proved that there are different smooth structures on four-dimensional real space, opening up questions about the relationships between algebraic and topological properties in the fourth dimension. He was awarded the Fields Medal in 1986 for this result and is professor of mathematics at Imperial College in London.

Douady, Adrien (b. 1935) French mathematician who has extensively studied the MANDELBROT SET. The rabbit-shaped FRACTAL that he studied is named Douady's rabbit after him. He is a professor at the École Normale Supérieure in Paris.

Douglas, Jesse (1897–1965) American mathematician who independently solved PLATEAU'S PROBLEM and made other

contributions to the study of MINIMAL SURFACES. He was
awarded the Fields Medal in 1936.

Dupin, François Pierre Charles (1784–1873) French naval engineer
who made his famous discovery of the Dupin's CYCLIDES as a
student of MONGE at the École Polytechnique in Paris. He made
other contributions to DIFFERENTIAL GEOMETRY, including the
Dupin indicatrix, while pursuing a career in the navy,
eventually becoming the French minister of maritime affairs.

Dürer, Albrecht (1471–1528) German artist who was interested in the
use of mathematics as a foundation for his art. While traveling
in Italy to learn more about art, he became acquainted with the
theory and practice of PERSPECTIVE and proportion. After his
second trip to Italy, Dürer wrote a four-volume work, the first
German books on mathematics. These books described the

From *About the Art of
Measurement* by Albercht
Dürer (Dover)

construction of curves, including the SPIRAL OF ARCHIMEDES, the LOGARITHMIC SPIRAL, and the CONCHOID, discussed the PLATONIC SOLIDS and the SEMIREGULAR POLYHEDRA, and gave an introduction to the theory of PERSPECTIVE. Dürer also invented NETS to describe the structure of polyhedra.

Eratosthenes (c. 276–c. 195 B.C.E.) Greek mathematician, astronomer, poet, and athlete who made a remarkably accurate calculation of the circumference of the Earth and created the sieve of Eratosthenes, a way to determine which numbers are prime. Born in Cyrene, now part of Libya, Eratosthenes studied in Athens and then went to Alexandria, where he spent the rest of his life.

Erdös, Paul (1913–96) Itinerant mathematician who was born and educated in Hungary but then traveled the world, seeking out mathematicians and mathematics problems. His work is remarkable for its depth and breadth, covering the areas of NUMBER THEORY, combinatorics, GRAPH THEORY, and discrete mathematics.

Escher, M. C. (1898–1970) Dutch artist who made frequent use of TILINGS, spheres, POLYHEDRA, HYPERBOLIC GEOMETRY, and other mathematical ideas in his work. Escher also made important mathematical discoveries and was the first to study and classify COLOR SYMMETRIES.

Euclid (c. 330–c. 270 B.C.E.) Greek mathematician whose name is synonymous with the study of geometry. He first studied at PLATO's Academy in Athens and then went to Alexandria around 300 B.C.E., where he wrote the *Elements* and other works on the CONIC SECTIONS, SPHERICAL GEOMETRY, PERSPECTIVE, and optics. The *Elements* included all the fundamental theorems in plane geometry, NUMBER THEORY, and SOLID GEOMETRY that were known at the time. The original contributions of Euclid presented in the *Elements* include a theory of parallel lines, a proof of the Pythagorean theorem, a proof of the infinitude of prime numbers, and the Euclidean algorithm for finding the greatest common denominator of two numbers. Because of its clarity, comprehensiveness, and logical sequence, it has become the standard for mathematical exposition. Euclid is famous for telling the ruler Ptolemy that there is "no royal road to geometry."

Eudoxus of Cnidus (c. 391–338 B.C.E.) Philosopher, physician, mathematician, and astronomer who traveled widely in the ancient world before eventually returning to his native city, Cnidus, in what is today Turkey. His theory of PROPORTIONS was the first step in the mathematical understanding of the REAL NUMBERS and resolved questions about the nature of INCOMMENSURABLES. His method of exhaustion was widely used to find the area and volume of CURVILINEAR figures.

Euler, Leonhard (1707–83) Swiss-born mathematician who made discoveries in all areas of mathematics and is regarded as one of the greatest mathematicians of all time. Born in Basel, where he was a friend of the Bernoulli family, Euler spent most of his life in Saint Petersburg at the court of Catherine the Great. His analysis of the Konigsberg Bridge Problem marked the beginning of TOPOLOGY and GRAPH THEORY. His other contributions to geometry include the EULER NUMBER, the EULER LINE, and Euler's formula for normal curvature.

Fano, Gino (1871–1952) Italian-born mathematician who studied in Göttingen under FELIX KLEIN. After his studies, he returned to Italy, where he taught for more than 40 years before moving to Switzerland. Fano was a pioneer in the study of FINITE GEOMETRY and discovered the first finite projective geometry, which now bears his name.

Fatou, Pierre (1878–1929) French mathematician who studied at the École Normale Supérieure in Paris and was awarded a doctoral degree in 1906 for results in the theory of complex functions. His work in the iterations of complex functions marked the beginning of the study of FRACTALS.

Federov, Efgraf Stepanovich (1853–1919) Russian crystallographer who independently discovered the 230 CRYSTALLOGRAPHIC GROUPS.

Feigenbaum, Mitchell Jay (b. 1944) American physicist who discovered the FEIGENBAUM NUMBER and its role in CHAOS theory. He received a doctoral degree in physics from the Massachusetts Institute of Technology (MIT) and then joined

Leonhard Euler (Image Select/Art Resource, NY)

the staff at Los Alamos National Laboratory. Feigenbaum is now a professor of physics at Rockefeller University.

Ferguson, Helaman (b. 1940) American mathematician and sculptor who has used his knowledge of geometry and TOPOLOGY to create beautiful and inspiring sculptures. His work helps the viewer understand the nature of abstract topological concepts and is displayed at many mathematical institutions throughout the country.

Fermat, Pierre de (1601–65) French judge and amateur mathematician who made significant and lasting contributions to NUMBER THEORY. He independently developed ANALYTIC GEOMETRY and the beginnings of CALCULUS. Along with BLAISE PASCAL, he founded the study of probability. His study of optics led to Fermat's principle of least time: A ray of light going from one point to another takes the path that requires the least time.

Feuerbach, Karl Wilhelm (1800–34) German mathematician who discovered the NINE-POINT CIRCLE when a student 22 years of age. He continued to work in geometry until forced to retire because of poor health.

Fibonacci (1180–1240) Italian merchant who contributed to many areas of mathematics. Leonardo Pisano or Leonardo of Pisa, better known as Fibonacci, discovered the Fibonacci series, 1, 1, 2, 3, 5, 8, . . . , which is used to model natural growth and has many deep mathematical properties.

Fourier, Joseph (1768–1830) French mathematician and scientist who introduced Fourier series, which are series of sines and cosines that sum to a specific periodic function, and made contributions to the theory of heat and linear differential equations. Fourier accompanied Napoleon to Egypt, where he had an opportunity to study Egyptian antiquities.

della Francesca, Piero (c. 1420–92) Italian Renaissance painter and mathematician who refined and furthered the use of the checkerboard in PERSPECTIVE geometry.

Freedman, Michael H. (b. 1951) American mathematician who was awarded the Fields Medal in 1986 for his solution of the POINCARÉ CONJECTURE for four-dimensional manifolds.

Pierre de Fermat (Getty)

Michael Freedman (MSR)

Freedman received his doctoral degree from Princeton University in 1973 and is now a researcher in the Theory Group at Microsoft Research, where he is working on quantum computation.

Fuller, Buckminster (1895–1983) American engineer and inventor who made extensive use of geometry in his work, which included the development of the GEODESIC DOME.

Galois, Évariste (1811–32) French algebraist who was the first to formalize the concept of SYMMETRY in terms of a GROUP of TRANSFORMATIONS. He was killed in a duel at the age of 20.

Garfield, James (1831–81) The 20th president of the United States. While a congressman in 1876, he found a new proof of the Pythagorean theorem that was based on the construction of a trapezoid containing a right triangle.

Gauss, Carl Friedrich (1777–1855) German mathematician who made great contributions to all areas of mathematics and was the most commanding figure of 19th-century mathematics. He helped found DIFFERENTIAL GEOMETRY with his understanding of CURVATURE and GEODESICS. Independently of Lobachevsky and Bolyai, he discovered NON-EUCLIDEAN GEOMETRY. Gauss gave the first proof of the fundamental theorem of algebra, which states that any polynomial equation with complex coefficients has a complex root.

Carl Friedrich Gauss
(The Granger Collection, New York)

Gelfand, Israil Moiseevic (b. 1913) Ukrainian mathematician who has been interested in functional analysis, algebra, and DIFFERENTIAL EQUATIONS, as well as applications of mathematics to cell biology. He immigrated to the United States in 1990 and received a MacArthur fellowship in 1994. He is now professor of mathematical methods in biology at Moscow State University.

Gergonne, Joseph Diaz (1771–1859) French military officer who devoted his life to mathematics. He gave an elegant solution to the PROBLEM OF APOLLONIUS and, along with PONCELET, discovered the principle of DUALITY in PROJECTIVE GEOMETRY. The GERGONNE POINT and GERGONNE TRIANGLE are named after him.

Gibbs, Josiah Willard (1839–1903) American physicist who made major contributions to thermodynamics, celestial mechanics, electromagnetism, and statistical mechanics. To further his work in physics, he developed vector calculus. He received the first doctorate in engineering to be awarded in the United States.

Girard, Albert (1595–1632) French-born Dutch mathematician and engineer who wrote a treatise on TRIGONOMETRY that introduced the abbreviations *sin, cos,* and *tan.* He computed formulas for the area of a SPHERICAL TRIANGLE.

Gordon, Carolyn (b. 1950) American mathematician who works in RIEMANN GEOMETRY and LIE GROUPS. In 1991, she helped to construct two different drums that sound the same when struck. She is professor of mathematics at Dartmouth College and past president of the Association for Women in Mathematics.

Graham, Ronald (b. 1935) American mathematician who has made significant contributions to combinatorics, GRAPH THEORY, NUMBER THEORY, COMBINATORIAL GEOMETRY, and the theory of algorithms. He received his doctorate from University of California at Berkeley and is now chief scientist at the California Institute for Telecommunication and Information Technology at the University of California at San Diego.

Grassmann, Hermann Günther (1809–77) German schoolteacher and scholar who received recognition for his mathematical work only after his death. His mathematical innovations included recursion, VECTOR SPACES, INNER PRODUCTS, and HIGHER-DIMENSIONAL geometry.

Grothendieck, Alexander (b. 1928) German-born French mathematician who made fundamental contributions to ALGEBRAIC GEOMETRY, TOPOLOGY, and abstract algebra. As a child he was taken by his mother from Germany to France, where he was able to receive an education during World War II. Noted for the power and abstraction of his work, Grothendieck was awarded the Fields Medal in 1966 for his work in algebraic geometry.

Grünbaum, Branko (b. 1929) American mathematician who was born in Yugoslavia, received his doctorate from Hebrew University and is now is professor emeritus at the University of Washington. He is a leading researcher in tilings, spatial patterns, polyhedra, polytopes, convexity, COMBINATORIAL GEOMETRY, and GRAPH THEORY.

Haken, Wolfgang (b. 1928) American mathematician who solved the FOUR-COLOR PROBLEM together with his colleague KENNETH APPEL at the University of Illinois. Their proof required more than 1,000 hours of computer calculations. Haken has also given an algorithm for determining whether or not a given KNOT DIAGRAM represents the UNKNOT.

Hales, Thomas (b. 1958) American mathematician who proved two important theorems, that the KEPLER CONJECTURE about the densest PACKING of spheres in space is true, and that the most efficient partition of the plane into equal areas is the regular hexagonal honeycomb. He is a professor of mathematics at the University of Pittsburgh.

Hamilton, William Rowan (1805–65) Irish mathematician and physicist who was an inventor of LINEAR ALGEBRA and vector calculus. A child prodigy, Hamilton had mastered 12 foreign languages by the age of 12. He developed the QUATERNION number system as a generalization of the complex numbers and was the Astronomer-Royal of Ireland.

Hausdorff, Felix (1868–1942) German mathematician who studied ANALYSIS, TOPOLOGY, and SET THEORY. He introduced the concepts of partially ordered set, METRIC SPACES, and HAUSDORFF DIMENSION.

Heawood, P. J. (1861–1955) English mathematician who is noted for his work in GRAPH THEORY and on the FOUR-COLOR PROBLEM.

Heesch, Heinrich (1906–1995) German mathematician who studied REGULAR TILINGS of the plane and the SYMMETRIES of colored tilings. He classified the 28 types of asymmetric tiles that can be used to create ISOHEDRAL TILINGS.

Helly, Eduard (1884–1943) Austrian-born mathematician who studied in Germany after completing his doctorate in Vienna. Helly's promising career in mathematics was interrupted by World War I, after which he immigrated to the United States and was able to resume his study of mathematics. He proved HELLY'S THEOREM about CONVEX SETS in 1923 and made important contributions to ANALYSIS.

Helmholtz, Hermann von (1821–94) A German mathematician, naturalist, and physician who worked in DIFFERENTIAL GEOMETRY and initiated the study of the INTRINSIC geometry of a surface.

Heron (c. 62) Greek mathematician and scientist residing in Alexandria who contributed to surveying, mechanics, pneumatics, and the study of mirrors. He developed numerical methods for computing square roots and cube roots and the Heron formula for the area of a triangle in terms of its sides. He is also known as Hero of Alexandria.

Hess, Edmund (1843–1963) German mathematician who studied POLYTOPES in depth and discovered the 10 regular STAR POLYTOPES.

Hessel, Johann (1796–1872) German crystallographer who determined the 32 CRYSTAL CLASSES and discovered the CRYSTALLOGRAPHIC RESTRICTION in three-dimensional space. He was a professor of mineralogy and mining technology at the University of Marburg.

Hilbert, David (1862–1943) German mathematician; one of the greatest mathematicians of all time. He worked in many areas, including geometry, algebra, NUMBER THEORY, ANALYSIS, the CALCULUS OF VARIATIONS, theoretical physics, and the foundations of mathematics. He developed a new system of axioms for Euclidean geometry that resolved problems that had been found in Euclid's *Elements*. Hilbert championed the axiomatic approach to mathematics and sought to strengthen the logical foundations of mathematics. In 1900, he gave a list of 23 problems that provided direction to 20th-century mathematics.

Hipparchus (c. 175–125 B.C.E.) Greek astronomer and mathematician who developed a simple form of TRIGONOMETRY, using the ratio of a chord to the diameter of a circle rather than the ratio

David Hilbert (Aufnahme von Fr. Schmidt, Göttingen, courtesy AIP Emilio Segrè Visual Archives, Lardé Collection)

of the sides of a right triangle, to assist in his astronomical calculations. He also discovered the precession of the equinoxes, which is due to the difference between the sidereal year and the solar year.

Hippias of Elis (c. 400 B.C.E.) Greek mathematician who invented the quadratrix, a curve that can be used to trisect an angle.

Hippocrates of Chios (c. 460–380 B.C.E.) Greek mathematician who computed the area of a LUNE. He also wrote a comprehensive textbook on geometry that has since been lost. Born on the island of Chios in the Aegean Sea, Hippocrates studied and taught mathematics in Athens.

de la Hire, Philippe (1640–1718) French geometer who wrote two books about the conic sections in the context of PROJECTIVE GEOMETRY. Originally an artist, de la Hire became interested in geometry while studying perspective during a trip to Italy. After returning to France, he was appointed to the chair of mathematics at the Collège Royale, where he continued his study of geometry.

Hopf, Heinz (1884–1971) German mathematician who worked in ALGEBRAIC TOPOLOGY and HOMOLOGY and created the algebraic structures known as Hopf algebras, which are used in algebraic topology and theoretical physics. After serving as an officer in World War I, Hopf completed his studies in Germany and traveled to France and the United States before settling at the Federal Institute of Technology in Zurich, Switzerland.

Hubbard, John (b. 1945) American mathematician who is a specialist in DYNAMICAL SYSTEMS and properties of the MANDELBROT SET. He received his doctorate from the University of Michigan and is now professor of mathematics at Cornell University.

Huntington, Edward V. (1874–1952) American mathematician who was interested in the axiomatic foundations of mathematics and developed several systems of axioms, including one for geometry.

Hurwitz, Adolf (1859–1919) German mathematician who was interested in NUMBER THEORY, COMPLEX ANALYSIS, and

RIEMANN SURFACES. Hurwitz was a student of FELIX KLEIN and teacher of DAVID HILBERT.

Ibn Qurra, Thābit (d. 901) Syrian mathematician at the court in Baghdad who translated works by EUCLID, ARCHIMEDES, and APOLLONIUS into Arabic and made original contributions to plane and SOLID GEOMETRY, TRIGONOMETRY, and algebra. He translated many classical Greek works into Arabic and wrote commentaries on them.

Ibn Yūnus (d. 1009) Egyptian astronomer who developed an interpolation procedure for calculating sines at half-degree intervals and compiled many numerical tables useful for astronomers.

Jackiw, Nick (b. 1966) American software designer who created the Geometer's Sketchpad dynamic geometry software and continues to refine it. Jackiw studied English literature and computer science at Swarthmore College before becoming interested in computer geometry.

Jacobi, Carl (1804–51) German mathematician who did research in many areas, including NUMBER THEORY, elliptic functions, DIFFERENTIAL EQUATIONS, and DIFFERENTIAL GEOMETRY. He developed the DETERMINANT now called the JACOBIAN.

Jones, Vaughn F. R. (b. 1952) American mathematician who was born in New Zealand and completed his graduate work in Switzerland. Jones was awarded the Fields Medal in 1990 for his work in KNOT THEORY and the discovery of the JONES POLYNOMIAL, a knot invariant that has connections with many different areas of mathematics and physics. He is now professor of mathematics at the University of California at Berkeley.

Jordan, Camille (1838–1922) French mathematician who made many contributions to algebra as well as geometry and calculus for higher-dimensional spaces. Jordan was the first mathematician to recognize the importance of the work of his compatriot ÉVARISTE GALOIS and wrote a monograph on the work of Galois to present it to other mathematicians. JORDAN CURVES are named after him.

Vaughn F. R. Jones
(UC Berkeley)

Julia, Gaston (1893–1978) French hero of World War I who worked on DYNAMICAL SYSTEMS while recovering from his wounds in a hospital. Julia's discovery of the important properties of JULIA SETS were appreciated only after BENOIT MANDELBROT displayed Julia sets on the computer and showed their relevance to the study of FRACTALS.

Kātyāyana (c. 200 B.C.E.) Vedic seer who wrote one of the main Sulba Sutras, which are the oldest known writings on geometry. His work includes approximate squaring of the circle and computation of a value for $\sqrt{2}$ that is accurate to five decimal places.

Kauffman, Louis (b. 1945) American mathematician who discovered the KAUFFMAN POLYNOMIAL, an important knot invariant. He has made numerous other contributions to KNOT THEORY, applications of knot theory to statistical mechanics, quantum theory, abstract algebra, combinatorics, and the foundations of mathematics. He received his doctorate from Princeton University in 1972 and is professor of mathematics at the University of Illinois at Chicago.

Kepler, Johannes (1571–1630) German astronomer and cosmologist who discovered the laws governing the motion of the planets around the Sun. An accomplished and creative geometer, Kepler based his astronomical work on his deep understanding of the Platonic solids and the conic sections. Kepler was the first to study the REGULAR and ARCHIMEDEAN TILINGS and show their connection to the regular polyhedra. He discovered all 11 Archimedean tilings and proved there were no others. He discovered several regular polyhedra and appears to be the first mathematician interested in sphere PACKINGS.

Johannes Kepler
(Giraudon/Art Resource, NY)

Khayyam, Omar (c. 1048–c. 1131) Persian astronomer, mathematician, poet, and philosopher who made contributions to arithmetic, algebra, and geometry. He solved the cubic equation by considering the solutions to be the intersection of conic sections and investigated the FIFTH POSTULATE of EUCLID.

Klee, Victor (b. 1925) American mathematician who has contributed to a wide range of areas of mathematics, including CONVEX SETS, mathematical programming, combinatorics, the design and analysis of algorithms, and POINT SET TOPOLOGY. He received his doctorate from the University of Virginia in 1949 and is now professor emeritus at the University of Washington.

Klein, Felix (1849–1925) German mathematician who had a lasting impact on all areas of geometry. In 1872, at the age of 23, Klein presented his ERLANGER PROGRAM, a way to unify different geometric theories using geometric TRANSFORMATIONS and their invariants. He later went on to make major contributions to PROJECTIVE GEOMETRY, TOPOLOGY, and ANALYSIS, in addition to continuing his work in transformation geometry. The KLEIN BOTTLE, a one-sided nonorientable surface that cannot exist in three-dimensional space, is named after him.

von Koch, Nils Fabian Helge (1870–1924) Swedish mathematician who is noted for creating the KOCH SNOWFLAKE to demonstrate the existence of a curve that is continuous but nowhere smooth. He studied at Stockholm University and later became a professor there.

Kolmogorov, Andrei Nikolaevich (1903–87) Russian mathematician who did research in probability, TOPOLOGY, ALGEBRAIC TOPOLOGY, functional analysis, DYNAMICAL SYSTEMS, and the foundations of geometry.

Kovalevskaya, Sonya Vasilievna (1850–91) Russian mathematician who made outstanding contributions to PARTIAL DIFFERENTIAL EQUATIONS and ANALYSIS. She studied in Germany under WEIERSTRASS and later accepted an academic position in Sweden.

Kuperberg, Krystyna (b. 1944) Polish-born American mathematician who studies TOPOLOGY, DISCRETE GEOMETRY, and DYNAMICAL SYSTEMS. She is a professor of mathematics at Auburn University.

Kuratowski, Kazimierz (1896–1980) Polish mathematician who was a leader of the Polish mathematical community and a world-renowned topologist. He worked on the foundations of POINT SET TOPOLOGY and developed the definition of TOPOLOGICAL

Felix Klein (Aufnahme von Fr. Struckmeyer, Göttingen, courtesy AIP Emilio Segrè Visual Archives, Lardé Collection)

SPACE that is now used. He also studied METRIC SPACES, COMPACT spaces, and GRAPH THEORY.

Laborde, Jean-Marie (b. 1945) French mathematician and computer scientist who led the development of the geometry software Cabri. He is research director of the Centre National de la Recherche Scientifique and head of the Cabri Geometry Project at the University of Grenoble.

Lagrange, Joseph Louis (1736–1813) Italian mathematician of French heritage who received great recognition for his work in DIFFERENTIAL EQUATIONS, algebra, NUMBER THEORY, and mechanics. Born in Turin, he became a professor of mathematics there at the age of 19, then spent 20 years in Berlin at the court of Frederick the Great before going to Paris, where he spent the rest of his life.

Lambert, Johann Heinrich (1728–77) German mathematician who proved that pi is IRRATIONAL. Of international renown, Lambert studied EUCLID'S PARALLEL POSTULATE and its negation, laying the groundwork for ELLIPTIC GEOMETRY. He also wrote a definitive work on the mathematical theory of PERSPECTIVE.

Lamé, Gabriel (1795–1870) French engineer who developed the theory of CURVILINEAR COORDINATES while working on the conductance of heat. Lamé worked on many problems with mathematical aspects and was considered to be the leading mathematician in France during his lifetime.

Laplace, Pierre Simon de (1749–1827) French astronomer and mathematician who is noted for writing *Celestial Mechanics,* which uses mathematics to describe the motion of the moon and planets. He also made advances in DIFFERENTIAL EQUATIONS and probability.

Lebesgue, Henri (1875–1941) French mathematician who generalized concepts of CALCULUS to make them more widely applicable. He also studied the CALCULUS OF VARIATIONS, the theory of SURFACE AREAS, and DIMENSION THEORY.

Leech, John (1926–92) British computer scientist and mathematician who discovered the Leech lattice, a LATTICE corresponding to a sphere packing in 24-dimensional space. He also studied NUMBER THEORY, geometry and combinatorial group theory.

Lefschetz, Solomon (1884–1972) Russian-born American mathematician who began his career as an engineer but switched to mathematics in 1910 after the loss of both hands in an industrial accident. He was a leader in the fields of ALGEBRAIC GEOMETRY, ALGEBRAIC TOPOLOGY, and DYNAMICAL SYSTEMS. He spent most of his mathematical career at Princeton University.

Legendre, Adrien Marie (1752–1833) French mathematician who studied NUMBER THEORY, ELLIPTIC INTEGRALS, and geometry. His *Eléments de géométrie* was influential in the reform of secondary mathematics education in France. He gave the first proof of EULER'S FORMULA for polyhedra.

Leibniz, Gottfried Wilhelm (1646–1716) German philosopher and mathematician who independently developed CALCULUS. Leibniz studied law in Germany and became interested in mathematics while he was visiting France on a diplomatic mission. He introduced symbolic logic, constructed a mechanical calculator, and worked on DETERMINANTS, COMPLEX NUMBERS, binary numbers, and combinatorics.

Lemoine, Émile (1840–1912) Lemoine studied at the École Polytechnique in Paris and became a civil engineer, amateur mathematician, and musician. He discovered that the SYMMEDIANS of a triangle meet at a point, now called the Lemoine point.

Leonardo da Vinci (1452–1519) Italian Renaissance painter, inventor, scientist, and engineer who was also a mathematician. In studying the designs of symmetric buildings, he discovered that the SYMMETRIES of a finite design must be either CYCLIC or DIHEDRAL.

Levi-Civita, Tullio (1873–1941) Italian mathematician who worked in DIFFERENTIAL GEOMETRY, contributing to the development to tensors, which are important in the theory of general relativity. Levi-Civita studied at the University of Padua and then taught there and at the University of Rome.

Gottfried Wilhelm Leibniz
(Image Select/Art Resource, NY)

Li Ye (1192–1279) Chinese mathematician who wrote the *Sea Mirror of Circle Measurements,* in which he used quadratic and cubic equations to solve problems of finding the diameters of circles that satisfy various given conditions.

Lie, Sophus (1842–99) Norwegian mathematician who developed LIE GROUPS and Lie algebras and made many other brilliant contributions to geometry and PARTIAL DIFFERENTIAL EQUATIONS. Lie became close friends with FELIX KLEIN during his student years in Berlin. Once, while traveling through France, he was arrested as a spy because his mathematical work looked like secret code.

Lindemann, Ferdinand von (1852–1939) German mathematician who proved that pi is TRANSCENDENTAL. Lindemann was a student of FELIX KLEIN at Erlangen and then continued his studies in England and France before returning to Germany to teach.

Liu Hui (c. 260) Chinese mathematician who made contributions to algebra and geometry. He wrote the *Sea Island Mathematical Manual* and the *Commentary on the Nine Chapters,* where he provides explanation and justification of material in the *Nine Chapters* and gives one of the earliest known proofs of the Pythagorean theorem. He computed a value of 3.14 for pi and computed the volume of a TETRAHEDRON, the FRUSTUM of pyramid, and many other solids.

Lobachevsky, Nicolai (1792–1856) Russian mathematician who independently discovered HYPERBOLIC GEOMETRY and many of its properties. Lobachevsky served as rector of the University of Kazan, where he was a popular teacher and administrator.

Lorentz, Hendrik Antoon (1853–1928) Dutch physicist who worked on the FitzGerald-Lorentz contraction used in special relativity and developed a mathematical theory of the electron for which he received the Nobel Prize in physics in 1902. A student and later professor at Leiden University in the Netherlands, Lorentz was primarily interested in theoretical physics.

Lorenz, Edward (b. 1917) American meteorologist at MIT who discovered the BUTTERFLY EFFECT and the phenomenon of CHAOS by studying a mathematical model of a weather system

Sophus Lie (Scanpix Norway)

Nicolai Lobachevsky
(The Granger Collection,
New York)

on a primitive computer. As of 2002 he was professor emeritus at MIT.

Madhava of Sangamagramma (c. 1340–1425) Astronomer and mathematician of southern India who wrote many astronomical treatises. He gave power series expansions for pi and for the SINE, COSINE, and INVERSE TANGENT functions. He created a table of the sine function accurate to eight or nine decimal places.

Mandelbrot, Benoit (b. 1924) American mathematician who is famous for the discovery of FRACTALS and the MANDELBROT SET. He was born in Poland, moved to France at the age of 12, and finally moved to the United States in 1958, where he was an IBM fellow. As of 2002 he was professor of mathematics at Yale University.

Mascheroni, Lorenzo (1750–1800) Italian priest, poet, and mathematician who is best known for proving that all Euclidean constructions can be made with compass alone.

Menaechmus (c. 350 B.C.E.) Greek mathematician who discovered the conic sections. He was a student of EUDOXUS and later tutor to Alexander the Great.

Menelaus (c. 100 C.E.) Greek astronomer residing in Alexandria who made extensive use of TRIGONOMETRY in his astronomical calculations and established trigonometry and SPHERICAL TRIGONOMETRY as distinct branches of mathematics.

Milnor, John (b. 1931) American mathematician who has been influential in the development of DIFFERENTIAL TOPOLOGY, ALGEBRAIC TOPOLOGY, MORSE THEORY, KNOT THEORY, and DYNAMICAL SYSTEMS. Appointed to the faculty of Princeton University at the age of 23, Milnor was awarded the Fields Medal in 1962 for his discovery of the EXOTIC SPHERES. He is now professor of mathematics at the State University of New York (SUNY) at Stony Brook.

Minkowski, Hermann (1864–1909) Russian-born mathematician who was the founder of the GEOMETRY OF NUMBERS and established the geometric framework for special relativity.

Benoit Mandelbrot
(Michael Marsland/Yale University)

John Milnor (John Griffin/ SUNY SB)

Minkowski studied mathematics in Germany, where he became close friends with DAVID HILBERT.

Möbius, August Ferdinand (1790–1868) German astronomer and mathematician who spent most of his life in Leipzig, where he was the director of the observatory. His major contributions to geometry include HOMOGENEOUS COORDINATES, BARYCENTRIC COORDINATES, INVERSION, and the MÖBIUS STRIP.

Mohr, Georg (1640–97) Danish mathematician who proved that all compass and straightedge constructions can be done with compass alone. Mohr studied in Holland, France, and England.

Moivre, Abraham de (1667–1754) French-born English mathematician who developed the formula $(\cos \theta + i \sin \theta)^n = (\cos n\theta + i \sin n\theta)$ for COMPLEX NUMBERS. He also studied ANALYTIC GEOMETRY and probability.

Monge, Gaspard (1746–1818) French mathematician who was a founder of PROJECTIVE GEOMETRY and DIFFERENTIAL GEOMETRY. He helped establish the École Polytechnique in Paris and made important contributions to ALGEBRAIC GEOMETRY and the geometry of curves and surfaces. Born the son of a shopkeeper in a small town in Burgundy, Monge attended a French military school where he showed early promise with discoveries in DESCRIPTIVE GEOMETRY.

Gaspard Monge (Réunion des Musées Nationaux/Art Resource, NY)

Moore, Robert L. (1882–1974) American mathematician and educator who made contributions to POINT SET TOPOLOGY and the foundations of topology. He developed the innovative Moore Method for teaching research skills in mathematics.

Morgan, Frank (b. 1952) American mathematician who studies minimal surfaces and was a collaborator on the proof of the double bubble theorem in 2000. He received his doctorate from Princeton University in 1977 and is now professor of mathematics at Williams College.

Moufang, Ruth (1905–77) German mathematician who contributed to the algebraic study of projective planes and theoretical physics. Moufang received her doctorate in 1931 for a thesis in PROJECTIVE GEOMETRY and was an industrial mathematician before being joining the University of Frankfurt, where she

Frank Morgan (courtesy Frank Morgan)

Isaac Newton (AIP Emilio Segrè Visual Archives, Physics Today Collection)

Emmy Noether (The Granger Collection, NY)

was the first woman in Germany to be appointed as a full professor.

Mydorge, Claude (1585–1647) French lawyer who devoted most of his life to mathematics. He wrote books on optics, the CONIC SECTIONS, and recreational mathematics. He was a friend of RENÉ DESCARTES and made many optical instruments for him.

Newton, Isaac (1643–1727) British physicist and mathematician who developed CALCULUS in order to solve problems in physics. He discovered gravity and used his newly invented calculus to show that his theory of gravity had as a consequence the elliptic orbits of the planets.

Noether, Emmy (1882–1935) German algebraist who pioneered the modern abstract approach to the study of groups. She proved that every conservation law in physics is related to a symmetry of space and also worked on the general theory of relativity. She was the first to recognize the algebraic nature of certain topological structures, thus contributing to the development of algebraic topology. She spent the last one and a half years of her life at Bryn Mawr College in Bryn Mawr, Pennsylvania.

Novikov, Sergei Petrovich (b. 1938) Soviet topologist who has made important contributions to DIFFERENTIAL TOPOLOGY, DYNAMICAL SYSTEMS, ALGEBRAIC TOPOLOGY, mathematical physics, and GAUGE FIELD theories. He was awarded the Fields Medal in 1970. Since 1996 he has been professor at the University of Maryland at College Park.

Pacioli, Luca de (1445–1514) Italian mathematician and Franciscan friar who traveled throughout Italy studying and teaching mathematics. A student of LEON ALBERTI, Pacioli was a close friend of LEONARDO DA VINCI, who made the drawings for his book on the golden ratio, *Divina proportione (The Divine Proportion)*.

Pappus (c. 300–c. 350) Greek mathematician who taught at the school in Alexandria and was the last great geometer of antiquity. His discoveries include three theorems that are now part of PROJECTIVE GEOMETRY.

Parmenides (c. 515–c. 450 B.C.E.) Greek philosopher who promoted the use of abstract reasoning and the law of the excluded middle. He founded a philosophical school in the colony of Elea on the southwestern coast of Italy and was a teacher of ZENO.

Pascal, Blaise (1623–62) French philosopher and writer who was also a gifted mathematician and made contributions to probability, NUMBER THEORY, and geometry. His MYSTIC HEXAGRAM THEOREM, discovered when he was 16 years old, has become a cornerstone of PROJECTIVE GEOMETRY.

Pasch, Moritz (1843–1930) German mathematician who was interested in the foundations of geometry and proposed new axioms to supplement Euclid's work.

Peano, Giuseppe (1858–1932) Italian mathematician who is best known for his axiomatic description of the REAL NUMBERS in terms of SET THEORY and his discovery of space-filling curves. Peano studied and later taught at the University of Turin in Italy.

Blaise Pascal (Giraudon/ Art Resource, NY)

Peaucellier, Charles-Nicolas (1832–1913) French engineer who designed a linkage based on INVERSION that converts a circular motion into a straight line. He was a graduate of the École Polytechnique.

Penrose, Roger (b. 1931) British theoretical physicist who has made numerous original contributions to geometry, including the PENROSE TILES. His research interests include nonperiodic tilings, general relativity, and the foundations of quantum theory. Along with his father, Lionel Penrose, he created the Penrose staircase, a circular staircase that seems to go up forever, and the tribar, an impossible triangle, both of which were used by M. C. ESCHER in his work. Roger Penrose is professor emeritus at Oxford University.

Petters, Arlie (b. 1964) American mathematician born in Belize who is the leading researcher in gravitational lensing, using SINGULARITY theory and CAUSTICS to explain how light from distant quasars is bent as it passes massive objects on its path to the Earth. Petters received his doctorate from MIT in 1991

Arlie Petters (Les Todd)

and has been given numerous awards and grants. As of 2002 he was professor of mathematics at Duke University.

Pick, Georg (1859–1942) Austrian mathematician who worked in COMPLEX ANALYSIS, geometry, and DIFFERENTIAL GEOMETRY. He is most noted for PICK'S THEOREM.

Plateau, Joseph (1801–83) Belgian physicist who studied visual perception, optical illusions, and the molecular forces in thin soap films. He formulated the PLATEAU PROBLEM of determining whether there is a MINIMAL SURFACE whose boundary is a given JORDAN CURVE in three-dimensional space.

Plato (429–347 B.C.E.) Greek philosopher who had a lasting influence on mathematics. Plato promoted the study of pure mathematics and described the construction of the five REGULAR POLYHEDRA, now called the Platonic solids. He limited the use of geometrical tools to the compass and straightedge and emphasized rigorous proof.

Playfair, John (1748–1819) Scottish minister, scholar, and mathematician who helped simplify and standardize Euclidean geometry in his edition of the *Elements*. His version of Euclid's FIFTH POSTULATE is known as the Playfair axiom.

Plücker, Julius (1801–68) German mathematician who studied PROJECTIVE GEOMETRY and ANALYTIC GEOMETRY. Plücker was also a physicist and discovered cathode rays.

Poincaré, Henri (1854–1912) French mathematician who made important contributions to many areas of mathematics and physics. Regarded as the greatest mathematician in the world during the late 19th century, Poincaré constructed models for HYPERBOLIC GEOMETRY—the POINCARÉ HALF-PLANE and the POINCARÉ DISK—that showed hyperbolic geometry to be as consistent as Euclidean geometry. He was a founder of the modern theories of TOPOLOGY, ALGEBRAIC TOPOLOGY, and DYNAMICAL SYSTEMS. His conjecture about three-dimensional manifolds is still open.

Henri Poincaré (Réunion des Musées Nationaux/Art Resource, NY)

Poinsot, Louis (1777–1859) French mathematician who discovered the four regular STELLATED polyhedra, now known as the Kepler-Poinsot solids.

Pólya, George (1887–1985) American mathematician who worked in NUMBER THEORY, combinatorics, and probability. Pólya studied mathematics in his native Hungary and in Germany, Switzerland, and England before immigrating to the United States. He proved that there are only 17 possible WALLPAPER PATTERNS and is noted for his teaching of problem-solving techniques.

Poncelet, Jean-Victor (1788–1867) French mathematician who founded the study of PROJECTIVE GEOMETRY. A soldier in Napoleon's army, Poncelet was captured in 1812 by the Russians and, while a prisoner of war, developed projective geometry to pass the time. After returning to France, he published his results and continued to work in projective geometry.

Jean-Victor Poncelet
(Getty)

Pontryagin, Lev Semenovich (1908–88) Russian mathematician who studied ALGEBRAIC TOPOLOGY and TOPOLOGICAL GROUPS. He also made major contributions to DIFFERENTIAL EQUATIONS, control theory, and DYNAMICAL SYSTEMS. Blinded in an accident at the age of 14, Pontryagin was aided in his study of mathematics by his mother.

Preparata, Franco (b. 1935) Italian-born mathematician and computer scientist who is a pioneer of COMPUTATIONAL GEOMETRY. He also studies the theory of computation, the design and analysis of algorithms, and information and coding theory. He is a professor of computer science at the University of Illinois.

Proclus (c. 410–485) Greek mathematician who was head of the Academy in Athens and wrote commentaries on Euclid's *Elements* as well as the oldest existing history of geometry.

Ptolemy, Claudius (c. 100–c. 178) Alexandrian astronomer and mathematician who developed a model for the motion of the planets using EPICYCLES. He discovered the sum, difference, and half-angle formulas and the Pythagorean identities in TRIGONOMETRY. He also made a compilation, called the *Almagest* by the Arabs, of all that was known in astronomy at the time.

Pythagoras (SEF/Art
Resource, NY)

Pythagoras (c. 560–c. 480 B.C.E.) Greek mathematician, philosopher, teacher, and scholar who has played a great role in the development of mathematics. He traveled widely in Egypt and Babylon before moving to Croton, where he founded a religious and philosophical society. Pythagoras believed that the nature of the material world could only be understood through mathematics. He is credited with the discovery of the Pythagorean theorem and the proof that $\sqrt{2}$ is irrational.

Quillen, Daniel (b. 1940) American-born mathematician who has made foundational contributions to HOMOLOGY THEORY, HOMOTOPY THEORY, and COHOMOLOGY THEORY. He was awarded the Fields Medal in 1978 for work in algebraic K-theory. Quillen is professor of mathematics at Oxford University.

Regiomontanus (1436–76) German mathematician and astronomer whose given name was Johann Müller. He wrote on plane and SPHERICAL TRIANGLES and the solution of triangles using the law of sines. He also gave the law of sines and the law of cosines for spherical triangles.

Reuleaux, Franz (1829–1905) German engineer who is a founder of modern kinematics. He gave an analysis of the kinematic properties of the REULEAUX TRIANGLE and other curves of constant width.

Rham, Georges de (1903–90) Swiss mathematician who made many important contributions to TOPOLOGY and RIEMANN GEOMETRY. De Rham studied in Switzerland, France, and Germany, before joining the University of Lausanne in Switzerland.

Riccati, Jacopo Francesco (1676–1754) Italian mathematician who did work in hydraulic engineering, ANALYSIS, DIFFERENTIAL EQUATIONS, and DIFFERENTIAL GEOMETRY. He studied CYCLOIDAL PENDULUMS and solved the Riccati differential equation.

Riccati, Vincenzo (1707–75) Italian mathematician and engineer noted for work on the HYPERBOLIC SINE and HYPERBOLIC COSINE. He was the son of mathematician JACOPO FRANCESCO RICCATI and furthered his father's work on integration and differential equations.

Ricci-Curbasto, Gregorio (1853–1925) Italian mathematician who made contributions to mathematical physics, DIFFERENTIAL EQUATIONS, and DIFFERENTIAL GEOMETRY. His work led to the development of TENSOR analysis.

Riemann, Bernhard (1826–66) German mathematician who introduced many important ideas that continue to shape modern mathematics. A student and later professor of mathematics at the University of Göttingen, Riemann studied higher-dimensional curved spaces, DIFFERENTIAL GEOMETRY, and COMPLEX ANALYSIS.

de Roberval, Gilles Personne (1602–75) French mathematician who studied plane curves—including the SINE curve, the cycloid, and spirals—and their tangents. He found the area under an arch of a CYCLOID and collaborated with FERMAT and PASCAL.

Bernhard Riemann
(The Granger Collection, New York)

Saari, Donald (b. 1940) American mathematician who has used geometry to study the theory of elections and voting. He received his doctorate from Purdue University in 1967. Saari is a member of the National Academy of Sciences and is professor of mathematics at University of California at Irvine.

Saccheri, Girolamo (1667–1733) Italian Jesuit priest who made extensive studies of Euclid's FIFTH POSTULATE, laying the groundwork for the discovery of HYPERBOLIC GEOMETRY.

Salmon, George (1819–1904) Irish mathematician and theologian who was one of the first mathematicians to study LINEAR ALGEBRA. He was an outstanding teacher whose textbooks on ANALYTIC GEOMETRY and algebra were influential in the development of modern algebra.

Schattschneider, Doris (b. 1939) American algebraist who has studied tilings and their symmetries, the relationship of mathematics to art, and mathematical work of M. C. ESCHER. She received her doctorate from Yale University in 1966, was professor of mathematics at Moravian College, and is now retired.

Schläfli, Ludwig (1814–95) Swiss teacher and self-trained mathematician who discovered many of the HIGHER-DIMENSIONAL POLYTOPES and completely classified the regular polytopes in n-dimensional Euclidean space.

Doris Schattschneider
(courtesy Doris Schattschneider)

Schönflies, Arthur Moritz (1853–1928) German mathematician who independently discovered the 230 CRYSTALLOGRAPHIC GROUPS in 1891. He also made important contributions to SET THEORY, TOPOLOGY, PROJECTIVE GEOMETRY, and kinematics.

Schoute, Pieter Hendrik (1846–1913) Dutch mathematician who studied POLYTOPES and HIGHER-DIMENSIONAL Euclidean geometry. He was a professor of mathematics at Groningen, Holland.

Schur, Friedrich Heinrich (1856–1932) German algebraist who studied TRANSFORMATION theory and the axiomatic foundation of Euclidean geometry.

Schwarz, Hermann Amandus (1843–1921) German mathematician who studied MINIMAL SURFACES and discovered the Schwarz minimal surface. The Cauchy-Schwarz inequality is named for him. He worked on conformal mappings and was a student of WEIERSTRASS.

Seki Kowa (1642–1708) Japanese mathematician of a samurai warrior family who used DETERMINANTS to solve equations and discovered the BERNOULLI numbers before they were known in the West. He also worked with magic squares and Diophantine equations.

Senechal, Marjorie (b. 1939) American mathematician who studies mathematical CRYSTALLOGRAPHY, DISCRETE GEOMETRY, QUASICRYSTALS, and the history of science and technology. She received her doctorate from the Illinois Institute of Technology in 1965 and is a professor at Smith College.

Serre, Jean-Pierre (b. 1926) French topologist who was awarded the Fields Medal in 1954 for the computation of HOMOTOPY GROUPS of higher-dimensional spheres based on the structure of LOOP spaces. Serre has also done work in ALGEBRAIC GEOMETRY and the TOPOLOGY of complex MANIFOLDS. He retired from his position at the Collège de France in 1994 when he became an honorary professor.

Shnirelman, Lev Genrikhovich (1905–38) Russian mathematician who made contributions to NUMBER THEORY and used

Marjorie Senechal
(Stan Sherer)

topological methods to solve problems in the CALCULUS OF VARIATIONS.

Shubnikov, Alexei Vassilievich (1887–1970) Russian crystallographer who studied crystal growth and developed the concepts of ANTISYMMETRY and SIMILARITY symmetry. He was a student of GEORGE WULFF.

Sierpinski, Waclaw (1882–1969) Polish mathematician who made important contributions to the development of SET THEORY and POINT SET TOPOLOGY. He invented SIERPINSKI SPACE to illustrate the axioms of topology.

Simson, Robert (1687–1768) Scottish mathematician who studied and wrote about Euclidean geometry. The SIMSON LINE is named after him even though it was discovered by WILLIAM WALLACE.

Singer, Isadore (b. 1924) American mathematician who has made important contributions to DIFFERENTIAL GEOMETRY and global analysis. Jointly with MICHAEL ATIYAH, he proved the Atiyah-Singer index theorem, which connects the analytic and topological properties of MANIFOLDS. He is professor of mathematics at MIT.

Smale, Stephen (b. 1930) American mathematician who has done groundbreaking work in DIFFERENTIAL TOPOLOGY, where he proved the generalized POINCARÉ CONJECTURE for dimensions greater than four and that a two-dimensional sphere can be turned inside out smoothly in four-dimensional space, and in DYNAMICAL SYSTEMS, where he introduced the HORSESHOE MAP. He was awarded the Fields Medal in 1966 and is professor emeritus at University of California at Berkeley.

Sommerville, Duncan (1879–1934) English geometer who described the PSEUDORHOMBICUBOCTAHEDRON and was the first to classify the Cayley-Klein geometries, which are PROJECTIVE GEOMETRIES endowed with a METRIC defined by a QUADRATIC FORM.

Staudt, Karl von (1798–1867) German mathematician who made important contributions to the development of PROJECTIVE GEOMETRY. He showed how to construct a regular polygon with 17 sides using a compass and straightedge. A student of

Stephen Smale
(UC Berkeley 1996)

GAUSS in Göttingen, von Staudt later taught at Erlangen University in Bavaria.

Steenrod, Norman (1910–71) American mathematician who refined and clarified the concept of FIBER BUNDLE, one of the most widely used tools in DIFFERENTIAL TOPOLOGY and ALGEBRAIC TOPOLOGY. He was a professor of mathematics at Princeton University.

Steiner, Jacob (1796–1863) Swiss geometer who was influential in the development of PROJECTIVE GEOMETRY, HARMONIC DIVISION, and CONIC SECTIONS. Steiner was a shepherd with no opportunity for education until he was nearly 20 years old.

Steinitz, Ernst (1871–1928) German algebraist who made important contributions to the classification of polyhedra. (*See* POLYHEDRON.)

Stott, Alicia Boole (1860–1940) British mathematician who investigated the properties of higher-dimensional POLYTOPES and determined the sections of the regular four-dimensional polytopes. Stott was the daughter of the creator of Boolean algebra, George Boole.

Struik, Dirk Jan (1894–2000) Dutch-born American mathematician who studied DIFFERENTIAL GEOMETRY and the history of mathematics. He was professor of mathematics at MIT.

Sullivan, Dennis (b. 1941) American mathematician who studies ALGEBRAIC TOPOLOGY, DIFFERENTIAL TOPOLOGY, and their applications to theoretical physics. He is professor of mathematics at the City University of New York.

Sumners, DeWitt (b. 1941) American mathematician who has done extensive work in using KNOT THEORY to understand the structure of DNA. He is professor of mathematics at Florida State University.

Sylvester, James Joseph (1814–97) British mathematician who studied LINEAR ALGEBRA, MATRICES, and the theory of invariants. Sylvester spent 12 years in the United States.

Tait, Peter Guthrie (1831–1901) Scottish mathematician and physicist who was a founder of KNOT THEORY and classified all knots

DeWitt Sumners (William Langford)

with up to seven CROSSINGS. He also worked on the FOUR-COLOR PROBLEM and was a professor at the University of Edinburgh.

Taylor, Brook (1685–1731) English mathematician who contributed to the early development of CALCULUS and introduced integration by parts and the TAYLOR SERIES. He also wrote a book on PERSPECTIVE that included a general treatment of VANISHING POINTS.

Taylor, Jean (b. 1944) American mathematician who studies MINIMAL SURFACES and soap bubbles and has developed mathematical models of crystals and crystal growth. Taylor first studied chemistry but switched to mathematics while a graduate student at the University of California at Berkeley; as of 2002 she was professor of mathematics at Rutgers University.

Jean Taylor (AWM)

Thales of Miletus (c. 625–c. 547 B.C.E.) Greek mathematician and astronomer who introduced the idea of proof into mathematics and discovered some of the first theorems of Euclidean geometry.

Theaetetus (c. 417–369 B.C.E.) Greek geometer who was reported to have discovered the OCTAHEDRON and the ICOSAHEDRON. He was a contemporary of PLATO and a student of Socrates and THEODORUS OF CYRENE.

Theodorus of Cyrene (c. 465–c. 399 B.C.E.) Greek mathematician who proved, along with THEAETETUS, that the square roots of the nonsquares up to 17 are IRRATIONAL. He was a highly regarded teacher of PLATO.

Thales (Alinari/Art Resource, NY)

Thom, René (1923–2002) French mathematician who made major contributions to ALGEBRAIC TOPOLOGY and DIFFERENTIAL TOPOLOGY with his work on characteristic classes, FIBER BUNDLES, and transversality. He was the originator of CATASTROPHE THEORY and was awarded the Fields Medal in 1958 for developing COBORDISM THEORY. He was a professor at the Institut des Hautes Études Scientifique in France.

Thompson, D'Arcy Wentworth (1860–1948) Scottish biologist who is noted for his book *On Growth and Form*, which analyzes the geometric forms appearing in nature. He was professor of the University of St. Andrews in Scotland for 64 years.

Thurston, William (b. 1946) American mathematician who revolutionized the study of TOPOLOGY in two and three dimensions using HYPERBOLIC GEOMETRY. He has also done work in KNOT THEORY and FOLIATIONS. Thurston was awarded the Fields Medal in 1982 and joined the mathematics department at Cornell University in 2003.

Torricelli, Evangelista (1608–47) Italian mathematician who studied CYCLOIDS and the CONIC SECTIONS. He computed the ARC LENGTH of a LOGARITHMIC SPIRAL and the areas under an arch of the CYCLOID. He was a student of Galileo.

Tychonoff, Andrei Nikolaevich (1906–93) Russian topologist who worked in TOPOLOGY, functional analysis, DIFFERENTIAL EQUATIONS, and mathematical physics. He showed how to define a topology on the CARTESIAN PRODUCT of topological spaces.

Uhlenbeck, Karen (b. 1942) American mathematician who works with PARTIAL DIFFERENTIAL EQUATIONS and their applications to four-dimensional MANIFOLDS and GAUGE FIELD theory. Professor of mathematics at the University of Texas, she received a MacArthur fellowship in 1982 and is a member of the National Academy of Sciences.

Urysohn, Pavel (1898–1924) Russian topologist who studied DIMENSION in the context of understanding the nature of curves and surfaces.

Vandermonde, Alexandre-Théophile (1735–96) French mathematician and experimental scientist who initiated the study of the DETERMINANT as a function defined for matrices and gave some of its properties, including the fact that the determinant of a MATRIX with two identical rows or two identical columns is zero. Vandermonde was a musician and only began his work in mathematics at the age of 35.

Varignon, Pierre (1654–1722) French Jesuit priest who discovered the theorem that bears his name and used CALCULUS to solve problems in mechanics and fluid flow. Varignon began his

Karen Uhlenbeck (Marsha Miller/University of Texas)

mathematical career at the age of 30 when he encountered Euclid's *Elements* by chance.

Veblen, Oswald (1880–1960) American mathematician who did foundational work in PROJECTIVE GEOMETRY, TOPOLOGY and DIFFERENTIAL GEOMETRY. A native of Iowa, Veblen developed a system of axioms for Euclidean geometry for his doctoral dissertation at the University of Chicago.

von Vega, Georg Freiherr (1754–1802) Slovenian mathematician and military officer who calculated pi to 140 decimal places and compiled tables of LOGARITHMS and TRIGONOMETRIC functions.

Viète, François (1540–1603) French lawyer and mathematician who introduced symbolic algebra. He also studied TRIGONOMETRY and its applications in astronomy; he was the first in western Europe to use all six trigonometric functions.

Viviani, Vincenzo (1622–1703) Italian geometer, physicist, and engineer who was a student of TORRICELLI and Galileo. He measured the velocity of sound, determined the tangents of the CYCLOID, and studied the shape now known as the TEMPLE OF VIVIANI.

Voronoi, Georgi (1868–1908) Russian mathematician who created VORONOI DIAGRAMS and was interested in NUMBER THEORY and the GEOMETRY OF NUMBERS. Voronoi studied mathematics at the University of St. Petersburg in Russia and then was appointed professor of mathematics at Warsaw University in Poland.

Wallace, William (1768–1843) Scottish mathematician who discovered the SIMSON LINE. Wallace was a professor of mathematics at Edinburgh University.

Wallis, John (1616–1703) English mathematician who wrote about ANALYTIC GEOMETRY, the CONIC SECTIONS, and the history of mathematics. He was a major contributor to the early development of CALCULUS.

Watt, James (1736–1819) English engineer who invented the steam engine and designed a MECHANISM for drawing straight lines that is still used.

Weierstrass, Karl (1815–97) German mathematician who developed a rigorous foundation for CALCULUS. His ideas and methods have had a profound influence on modern mathematics, making it possible to handle LIMITS and the infinite nature of the REAL NUMBERS in a precise and systematic manner.

Wessel, Caspar (1745–1818) Norwegian surveyor and cartographer who worked in Denmark. He gave a geometric interpretation to the COMPLEX NUMBERS and their operations.

Weyl, Hermann (1885–1955) German-born American mathematician who was a student of DAVID HILBERT and received his doctorate from the University of Göttingen. He made important contributions to DIFFERENTIAL GEOMETRY, the general theory of relativity, TOPOLOGY, GROUP theory, ANALYSIS, and theoretical physics.

Whitehead, George (b. 1918) American topologist who worked in ALGEBRAIC TOPOLOGY and HOMOTOPY theory. He is a professor emeritus at MIT.

Whitehead, J. H. C. (1904–60) Mathematician who was born in India to English parents and received his doctorate from Princeton University. He did foundational work in DIFFERENTIAL GEOMETRY, DIFFERENTIAL TOPOLOGY, and ALGEBRAIC TOPOLOGY and was a professor of mathematics at Oxford University.

Whitney, Hassler (1907–89) American mathematician who worked in TOPOLOGY, DIFFERENTIAL TOPOLOGY, and GRAPH THEORY at Harvard University and the Institute for Advanced Study. He was awarded the National Medal of Science in 1975.

Whyburn, Gordon (1904–69) American mathematician who was a leader in POINT SET TOPOLOGY and analytic topology. He taught at the University of Virginia.

Witt, Jan de (1625–72) Dutch mathematician and statesman who wrote one of the first textbooks on ANALYTIC GEOMETRY, in which he introduced the modern treatment of the CONIC SECTIONS as a LOCUS of points satisfying certain properties and in terms of their focus and directrix. He also applied mathematics to the study of annuities and economics.

Witten, Edward (b. 1951) American mathematician and physicist who has made fundamental contributions in the areas of TOPOLOGY, theoretical physics, KNOT THEORY, SUPERSYMMETRY, and MORSE THEORY. Witten is a mathematical physicist at the Institute for Advanced Study and was awarded the Fields Medal in 1990.

Wulff, George (1863–1925) Ukrainian crystallographer who introduced mathematical ideas to model the forces that produce crystals.

Yang Hui (c. 1261–76) Chinese mathematician who wrote the seven-volume *Method of Computation of Yang Hui* in which he gives the formula for the solution of the general quadratic equation and formulas for finding areas and volumes. He also created many magic squares, some as large as order 10.

Yang, Chen Ning (b. 1922) Chinese-born American physicist who collaborated on the Yang-Mills GAUGE THEORY. He is a member of the Institute for Theoretical Physics at SUNY Stony Brook.

Yau, Shing-Tung (b. 1949) Chinese-born American mathematician who works in DIFFERENTIAL EQUATIONS, ALGEBRAIC GEOMETRY, MINIMAL SURFACES, and theoretical physics. Yau studied at Berkeley under S. S. CHERN and is now a professor of mathematics at Harvard University. He was awarded the Fields Medal in 1982 and is a member of the National Academy of Sciences.

Yorke, James (b. 1941) China-born American mathematician who first introduced the term *chaos* into the study of DYNAMICAL SYSTEMS. He is interested in the applications of dynamical systems to modeling the weather and the spread of infectious diseases. He is professor of mathematics and director of the Institute for Physical Science and Technology at the University of Maryland.

Young, John Wesley (1879–1932) American mathematician and mathematics educator who developed a system of POSTULATES for PROJECTIVE GEOMETRY with Oswald Veblen. He received his doctorate in mathematics from Cornell University and taught at Dartmouth College.

Edward Witten (IAS)

Shing-Tung Yau
(Kris Snibbe/Harvard News Office)

Zeeman, Christopher (b. 1925) English mathematician who has made contributions to TOPOLOGY, DYNAMICAL SYSTEMS, and CATASTROPHE THEORY. He invented the ZEEMAN CATASTROPHE MACHINE and was a founder of the mathematics department of the University of Warwick.

Zeno (c. 490–c. 425 B.C.E.) Greek philosopher and logician who proposed many paradoxes that brought to light difficulties in understanding space, time, motion, and relative motion. His most famous paradoxes are the Dichotomy, Achilles and the Tortoise, the Flying Arrow, and the Stadium. He was a student of PARMENIDES.

Zenodorus (c. 200–140 B.C.E.) Greek mathematician and astronomer who studied isometric figures, particularly the area of a figure with a fixed perimeter and the volume of a solid with fixed surface. He showed that the regular polygon has the greatest area of polygons with equal perimeter and an equal number of sides; that a circle has greater area than any regular polygon of the same perimeter; and that the sphere has the greatest surface area for a given volume.

Zhu Shijie (c. 1290–1303) Chinese mathematician and noted teacher who wrote the *Introduction to Mathematical Studies* and *Jade Mirror of the Four Elements,* which was devoted primarily to algebra and included PASCAL's triangle.

Zu Chongzhi (c. 429–500) Chinese astronomer and engineer who computed the value of 355/113 for pi, accurate to seven decimal places, and, together with his son Zu Geng, found the volume of a sphere.

SECTION THREE
CHRONOLOGY

c. 2600 B.C.E. ● The Great Pyramid at Giza is built by the Egyptians

c. 1850 B.C.E. ● An unknown Egyptian scribe records methods of calculating areas and volumes, including the volume of the frustum of a square pyramid on a papyrus. The Egyptians use a tool consisting of a plumb line and sight rod to make measurements in surveying and astronomy

c. 1800 B.C.E. ● A Mesopotamian scribe records a table of Pythagorean triples on a clay tablet

c. 1650 B.C.E. ● The Egyptian scribe A'hmosè copies a mathematical handbook that shows how to compute the area of a circle, the area of a triangle, and the steepness of a pyramid. He uses the value 3.1605 for pi

c. 1500 B.C.E. ● The Egyptians use a sundial to measure time

c. 600 B.C.E. ● The Sulba Sutras, which contain ancient Vedic knowledge about the building of ceremonial platforms, are written down in Sanskrit. They include the Pythagorean theorem, empirical procedures for making many geometric constructions, formulas for finding the areas of squares, rectangles, trapezoids, and circles, and formulas for finding the volume of prisms and cylinders

585 B.C.E. ● The Greek mathematician THALES predicts a solar eclipse using mathematical methods. Around this time, Thales also computes the height of the Great Pyramid in Egypt using similar triangles

c. 530 B.C.E. ● A tunnel is built straight through a mountain on the Greek island of Samos. Geometrical calculations are used to ensure that workers digging from both ends will meet each other in the middle

c. 510 B.C.E. ● PYTHAGORAS founds a school in Croton (later part of Italy) incorporating the belief that mathematics can elevate the human soul to union with the divine. The Pythagorean school, instrumental in the early development of geometry, studies FIGURATE NUMBERS, discovers the PYTHAGOREAN THEOREM, proves that $\sqrt{2}$ is irrational, and proves that the sum of the angles of a triangle is 180°

A section of the papyrus of A'hmosè (Art Resource, NY)

c. 387 B.C.E. ● PLATO founds the Academy in Athens, where mathematics was the main subject of instruction. The inscription on the gate to the Academy reads, "Let no one ignorant of geometry enter here"

386 B.C.E. ● An oracle tells the people of Delos that, to end a plague, they must double a cube-shaped altar of Apollo. This is the origin of the Delian problem

c. 350 B.C.E. ● The Chinese *Book of Crafts,* which gives practical applications of geometry and describes the measure of angles and arcs, and the more theoretical *Book of Master Mo,* which introduces the geometric concepts of point, line, surface, solid, and congruence, are written

c. 300 B.C.E. ● The Museum is founded in Alexandria by the ruler Ptolemy Soter. For more than 700 years, it will be a center of learning and home to many great mathematicians, including EUCLID, who compiles the *Elements,* a complete study of triangles, rectangles, circles, polygons, proportion, similarity, the theory of numbers, and solid geometry, including pyramids, cones, cylinders, and the five regular polyhedra. In the *Elements,* Euclid presents geometry as an axiomatic system, beginning with postulates and common notions that are assumed to be true and then deriving hundreds of theorems in a sequential (step by step) logical fashion

c. 240 B.C.E. ● The Greek mathematician Nicomedes invents the conchoid and uses it to trisect angles

c. 230 B.C.E. ● The Greek ERATOSTHENES applies geometric reasoning to measurements of the inclination of the Sun on the summer solstice to estimate the diameter of the Earth

212 B.C.E. ● Roman soldiers pillaging Syracuse kill ARCHIMEDES while he is contemplating a geometric problem. During his lifetime, Archimedes made major contributions to geometry, astronomy, engineering, and hydrostatics

c. 200 B.C.E. ● APOLLONIUS uses EPICYCLES and ECCENTRIC CIRCLES to model the motion of the planets. He also writes the *Conics,* a comprehensive treatise on the conic sections. It will become the standard source for properties of the conic sections and will be used by KEPLER, DESCARTES, and NEWTON

c. 150 B.C.E. ● Hipparchus uses geometry to compute the distance to the Sun and the Moon from the Earth

c. 130 B.C.E. ● Trigonometry begins with the use of chords by HIPPARCHUS

c. 100 B.C.E. ● *Chiu Chang Suan Shu (Nine Chapters on the Mathematical Arts)* is completed by Chinese scholars. It states the PYTHAGOREAN THEOREM and shows how to compute areas and volumes and solve linear and quadratic equations

c. 140 C.E. ● PTOLEMY writes the *Almagest,* a treatise on astronomy. It includes the first trigonometric tables and Ptolemy's theory of EPICYCLES and the EQUANT to model the motion of the planets with respect to the Earth

c. 470 ● The Chinese mathematician Tsu Chung Chih gives the value of 355/113 for pi, which is correct to six decimal places

628 ● BRAHMAGUPTA discovers a complete description of all rational triangles and gives a formula for the area of a cyclic quadrilateral in terms of the lengths of its sides

c. 1000 ● The Egyptian astronomer IBN YŪNUS draws up trigonometric tables for use in astronomical observations

c. 1240 ● The Alhambra is built in Granada, Spain. Its elaborate tilings will inspire artists and mathematicians for centuries

1248 ● The Chinese mathematician LI YE completes the *Tshe Yuan Hai Ching (Sea Mirror of Circle Measurements),* a work that includes methods of solving geometric problems involving circles

c. 1430 ● In the *Treatise on the Chord and Sine,* the Arabic astronomer al-Kashi introduces methods of using decimal fractions for the calculation of trigonometric tables

1435 ● LEON ALBERTI BATTISTA completes *Della pittura,* the first written account of the rules of PERSPECTIVE drawing, including rules for drawing a checkerboard floor in perspective. It will not, however, be printed until 1540

c. 1460 ● PIERO DELLA FRANCESCA writes *De prospettiva pingendi,* which gives techniques for drawing three-dimensional objects in PERSPECTIVE

1482 ● EUCLID's *Elements* is printed for the first time; it will become the most widely used mathematics textbook in history

1509 ● LUCA DE PACIOLI writes *De divina proportione,* a treatise on the golden section

1569 ● The Flemish geographer Gerardus Mercator makes the first map that used STEREOGRAPHIC PROJECTION of the Earth onto a cylinder that is then unrolled. Such maps will later be called Mercator maps

1570 ● Henry Billingsley completes the first translation into English of EUCLID's *Elements*

1603 ● The German astronomer Cristoph Scheiner invents the PANTOGRAPH

1609 ● JOHANNES KEPLER's *Astronomia Nova* describes his first two laws of planetary motion, that planetary orbits are ELLIPSES with the Sun at a focus and that the radius from a planet to the Sun sweeps out equal areas in equal times

1611 ● JOHANNES KEPLER proposes an explanation for the sixfold symmetry found in snowflakes that is based on the CUBIC CLOSE PACKING of spheres

1614 ● The Scottish landowner and inventor John Napier announces his invention of LOGARITHMS, which are first used to assist in computations involving trigonometric functions

1619 ● JOHANNES KEPLER's *Harmonice Mundi* is published. It is the first systematic mathematical study of TILINGS of the plane and their relationship to polyhedra and contains a picture of the 11 ARCHIMEDEAN TILINGS

1621 ● The British mathematician William Oughtred invents the slide rule, which uses LOGARITHMS to perform mathematical calculations.

● The Dutch scientist Willebrord Snell formulates the law of refraction for light passing from one medium to another, that the ratio of the sines of the angles of incidence and reflection is a constant depending on the two media

1634 ● Pierre Herigone first uses the symbol ∠ for an angle

1635 ● BONAVENTURA CAVALIERI introduces the principle that two solids with equal altitudes have the same volume if parallel cross sections at equal heights have equal areas. It will become known as Cavalieri's principle.

1636 ● GILLES PERSONNE DE ROBERVAL computes the area and the center of gravity of an arch of the CYCLOID

1637 ● RENÉ DESCARTES develops ANALYTIC GEOMETRY and also proves that the duplication of the cube cannot be done with compass and straightedge

1639 ● GIRARD DESARGUES publishes a treatise on PROJECTIVE GEOMETRY. Far ahead of its time, it will be ignored for almost 200 years

1641 ● Evangelista Torricelli discovers important properties of the CYCLOID

1666 ● Isaac Newton develops CALCULUS but does not publish his discovery until 1711

1672 ● GEORG MOHR proves that all Euclidean constructions that can be done with compass and straightedge can be done with compass alone

1674 ● GOTTFRIED WILHELM LEIBNIZ shows that pi is equal to the sum of the infinite SERIES $4\left(1 - \frac{1}{3} + \frac{1}{5} - \frac{1}{7} + \frac{1}{9} - \frac{1}{11} + \ldots\right)$

1675 ● GOTTFRIED WILHELM LEIBNIZ develops differential calculus

1678 ● GIOVANNI CEVA proves Ceva's theorem, which describes the conditions for three cevians of a triangle to be concurrent

1694 ● JACOB BERNOULLI introduces POLAR COORDINATES

1696 ● JACOB BERNOULLI and his brother JOHANN BERNOULLI independently solve the BRACHISTOCHRONE problem of determining the shape of a wire along which a frictionless bead will fall in the least time. The solution is the CYCLOID curve. Jacob Bernoulli's solution leads to the development of the CALCULUS OF VARIATIONS

1698 ● Samuel Reyher introduces the symbol ⌐ for a right angle

1731 ● ALEXIS CLAUDE CLAIRAUT is the first to use three-dimensional analytic geometry to study space curves

● Varignon's theorem, which says the MEDIANS of a quadrilateral form a parallelogram, is published

1733 ● GIROLAMO SACCHIERI publishes a book in which he attempts to prove Euclid's FIFTH POSTULATE. His results lay the foundations of what will become HYPERBOLIC GEOMETRY

1736 ● By solving the problem of the Konigsberg bridges, LEONHARD EULER at once creates the fields of TOPOLOGY and GRAPH THEORY. He also introduces the symbol π for the ratio of a circle's circumference to its diameter

1751 ● LEONHARD EULER discovers that the number of faces minus the number of edges plus the number of vertices for a simple closed polyhedron is two. It will become known as Euler's formula

1760 ● GEORGES BUFFON shows how to approximate the value of pi by repeatedly dropping a needle onto a plane ruled with parallel lines. If the length of the needle is l and the parallel lines are distance d apart, with $l < d$, the probability that a needle dropped at random will intersect one of the lines is $2l/\pi d$

1761 ● The German mathematician JOHANN LAMBERT proves that pi is irrational

1764 ● JEAN D'ALEMBERT suggests that time is a fourth dimension

1765 ● The Euler line is discovered by LEONHARD EULER

1775 ● Giulio Fagnano proposes and solves the problem of finding, for a given acute triangle, the inscribed triangle whose perimeter is as small as possible

1784 ● JAMES WATT, inventor of the steam engine, designs a LINKAGE consisting of four rods that can be used to construct a very close approximation to a straight line

1789 ● GEORG VON VEGA calculates pi to 140 places, a record that will stand for more than 50 years

1795 ● GASPARD MONGE publishes a treatise on descriptive geometry that signals a renewed interest in geometry

● JOHN PLAYFAIR'S edition of EUCLID's *Elements* is published, introducing the Playfair axiom, a simpler form of Euclid's FIFTH POSTULATE

1796 ● At the age of 19, CARL FRIEDRICH GAUSS proves that a regular polygon with *n* sides can be constructed with compass and straightedge if and only if the odd prime factors of *n* are distinct Fermat primes

1797 ● Independently of GEORG MOHR, LORENZO MASCHERONI proves that any construction that can be made with compass and straightedge can be made with compass alone

1799 ● The SIMSON LINE is discovered by WILLIAM WALLACE

1809 ● The French mathematician LOUIS POINSOT describes the structure of the GREAT DODECAHEDRON and the GREAT ICOSAHEDRON

1820 ● JEAN-VICTOR PONCELET and CHARLES BRIANCHON publish a proof of the existence of the NINE-POINT CIRCLE

1822 ● JEAN-VICTOR PONCELET's treatise on PROJECTIVE GEOMETRY is published and receives wide recognition and acceptance

● KARL WILHELM FEUERBACH discovers that the NINE-POINT CIRCLE of a triangle is tangent to the INSCRIBED CIRCLE and the three EXCIRCLES of the triangle

● MORITZ PASCH proposes a set of axioms for Euclidean geometry that is designed to complete the logical gaps in Euclid's axioms. One of these axioms, that a line entering a triangle at a vertex intersects the opposite side and a line intersecting one side of a triangle other than at a vertex also intersects exactly one other side of the triangle, is known as Pasch's axiom

● R. J. Haüy proposes the theory that crystals are composed of building blocks in the shape of polyhedra stacked together in a systematic way

1826 ● NICOLAI LOBACHEVSKY gives the first public lecture on HYPERBOLIC GEOMETRY at Kazan University, beginning the development of non-Euclidean geometry

1830 ● The finite GROUPS of ISOMETRIES of three-dimensional space are determined by JOHANN HESSEL

1831 ● HOMOGENEOUS COORDINATES for PROJECTIVE GEOMETRY are introduced by JULIUS PLÜCKER. More than a century later, they will be used by computers to render three-dimensional objects on a two-dimensional screen

1832 ● The French mathematician ÉVARISTE GALOIS invents GROUPS as a way to quantify and understand SYMMETRY

● The Swiss mathematician JACOB STEINER defines point conics and line conics as the appropriate way to study the conic sections in PROJECTIVE GEOMETRY

1837 ● Pierre Wantzel proves that the CONSTRUCTIBLE NUMBERS are the numbers that are formed from the integers through repeated applications of addition, multiplication, subtraction, division, and the extracting of square roots. In particular, this means that angles cannot be trisected with compass and straightedge

1838 ● Miquel's theorem, about the concurrency of the three MIQUEL CIRCLES of a triangle, is proved by Auguste Miquel

1840 ● C. L. Lehmus sends the statement of the Steiner-Lehmus theorem to JACOB STEINER, who subsequently proves it

1843 ● WILLIAM ROWAN HAMILTON discovers QUATERNIONS, a generalization of the complex numbers, while walking across a bridge in Dublin. Recognizing the importance of his discovery, he carves the formulas describing the properties of quaternions into the railing of the bridge

1844 ● HERMANN GRASSMANN introduces the concept of higher-dimensional VECTOR SPACES

1845 ● ARTHUR CAYLEY discovers the OCTONIANS, or Cayley numbers, which are an extension of the quaternions. The octonians are neither commutative nor associative

1850 ● AUGUSTE BRAVAIS discovers the 14 three-dimensional LATTICES

● PETER LEJEUNE DIRICHLET first studies VORONOI DIAGRAMS

1854 ● BERNHARD RIEMANN delivers a talk on the foundations of geometry at Göttingen University in which he lays out the principles of ELLIPTIC GEOMETRY and RIEMANN GEOMETRY

1858 ● ARTHUR CAYLEY develops matrices and the rules for MATRIX MULTIPLICATION

● The MÖBIUS STRIP is discovered independently by AUGUST FERDINAND MÖBIUS and Johann Benedict Listing

1865 ● LEONHARD EULER discovers the EULER LINE, passing through the circumcenter, the orthocenter, and the centroid of a triangle

1868 ● EUGENIO BELTRAMI shows that the surface of the PSEUDOSPHERE satisfies the axioms of HYPERBOLIC GEOMETRY, convincing the mathematical community of the value of non-Euclidean geometry

● French engineer Charles-Nicolas Peaucellier designs a linkage that can be used to construct a straight line. It consists of six rods and is based on INVERSION with respect to a circle

1870 ● CAMILLE JORDAN publishes a comprehensive exposition of the works of ÉVARISTE GALOIS on GROUPS, SYMMETRY, and the solvability of polynomial equations that first introduces these ideas to the mathematical community

1871 ● FELIX KLEIN gives the names *hyperbolic, parabolic,* and *elliptic* to the non-Euclidean geometry of LOBACHEVSKY and BOLYAI, Euclidean geometry, and Riemannian geometry, respectively

1872 ● FELIX KLEIN presents an address at the University of Erlangen in which he describes how to understand and classify different geometries according to their GROUPS of SYMMETRY TRANSFORMATIONS. This approach will later be called the ERLANGER PROGRAM

● The KLEIN BOTTLE is first constructed by FELIX KLEIN

1873 ● Charles Hermite proves that the real number *e,* used as the basis for NATURAL LOGARITHMS, is TRANSCENDENTAL

● RADIAN measure for angles is developed by mathematician Thomas Muir and physicist James T. Thompson in England

● SOPHUS LIE begins the development of the theory of LIE GROUPS

1878 ● ARTHUR CAYLEY presents the FOUR-COLOR PROBLEM to the London Mathematical Society. The problem has been stated earlier but will not be solved until 1976

1882 ● FERDINAND VON LINDEMANN proves that pi is a TRANSCENDENTAL NUMBER and that $\sqrt{\pi}$ is not a CONSTRUCTIBLE NUMBER, thereby proving that it is impossible to square the circle with compass and straightedge

● Hermann Schwarz proves that the sphere has the least surface area of all solids with a given volume

1885 ● The Russian crystallographer E. S. FEDEROV discovers the 230 three-dimensional crystallographic groups

1888 ● The definitive modern edition of EUCLID's *Elements* is published; it was prepared by Danish philologist Johan Ludvig Heiberg

1889 ● GIUSEPPE PEANO develops a new set of axioms for Euclidean geometry

1890 ● GIUSEPPE PEANO invents the PEANO CURVE, the first example of a space-filling curve

● PERCY J. HEAWOOD solves the FOUR-COLOR PROBLEM on the surface of a torus

● The American mathematician C. N. Little and the Scottish physicist PETER GUTHRIE TAIT publish diagrams of knots having up to 11 crossings; there are approximately 800 such knots

● The French mathematician HENRI POINCARÉ founds the study of dynamical systems by considering the qualitative behavior of all possible solutions to a differential equation rather than the specific behavior of one particular solution. In doing this, he is the first to discover the phenomenon of CHAOS in the solutions of a differential equation

1891 ● The German mathematician ARTHUR M. SCHOENFLIES independently discovers the 230 three-dimensional CRYSTALLOGRAPHIC GROUPS

● E. S. FEDEROV discovers the 17 different WALLPAPER GROUPS

1892 ● GINO FANO creates the first FINITE GEOMETRY

1895 ● HENRI POINCARÉ generalizes the Euler theorem for polyhedra to MANIFOLDS. His result is known as the EULER-POINCARÉ FORMULA. He also introduces the FUNDAMENTAL GROUP into the study of TOPOLOGY

1896 ● HERMANN MINKOWSKI proves the CONVEX BODY THEOREM

1897 ● Raoul Bricard proves that every plate and hinge FRAMEWORK shaped like an OCTAHEDRON is rigid, even though there are octahedral rod and joint frameworks that are not rigid

1899 ● GEORG PICK determines the area of a polygon whose vertices are LATTICE POINTS; his result is known as PICK'S THEOREM

● In his work *Foundations of Geometry,* DAVID HILBERT develops a set of axioms for Euclidean geometry that incorporates the concept of betweenness for points on a line

● MORLEY'S THEOREM is discovered by the British mathematician Frank Morley

1900 ● DAVID HILBERT addresses the International Congress of Mathematicians in Paris, outlining 23 problems that he feels will be important to 20th-century mathematicians. Of these, six are related to geometry

● MAX DEHN solves the third of Hilbert's problems by using the invariance of DIHEDRAL CONTENT to show that a regular tetrahedron is not EQUIDECOMPOSABLE with the cube

1902 ● The American mathematician F. R. Moulton discovers the Moulton plane, an AFFINE GEOMETRY in which the THEOREM OF DESARGUES is not valid

1903 ● HELGE VON KOCH creates the KOCH SNOWFLAKE as an example of a curve that is CONTINUOUS but nowhere DIFFERENTIABLE. It will later be studied as a simple example of a FRACTAL

1904 ● HENRI POINCARÉ states the conjecture that a three-dimensional MANIFOLD with the same HOMOTOPY type as a sphere must be TOPOLOGICALLY EQUIVALENT to a sphere. This conjecture will remain unproven, although the generalization to higher dimensions will be proved

● OSWALD VEBLEN develops a new system of axioms for Euclidean geometry

1906 ● Maurice Fréchet introduces abstract spaces and the concept of COMPACTNESS

● The study of FINITE PROJECTIVE GEOMETRIES is initiated by OSWALD VEBLEN and William Henry Bussey

1908 ● GEORGI VORONOI introduces VORONOI DIAGRAMS

1912 ● L. E. J. BROUWER proves the Brouwer fixed point theorem, which says that any continuous function from a disk to itself has a fixed point

1913 ● EDWARD V. HUNTINGTON develops a new system of axioms for Euclidean geometry.

1914 ● MAX DEHN proves that the left-hand TREFOIL and the right-hand trefoil are not equivalent

1915 ● Albert Einstein formulates his general theory of relativity, which implies that SPACETIME is a curved Riemannian space, rather than a flat Euclidean space as had been believed

1917 ● Mary Haseman lists all AMPHICHEIRAL KNOTS with 12 crossing

1919 ● Observation of a total solar eclipse by Sir Arthur Eddington confirms Albert Einstein's general theory of relativity

1920 ● HERMANN WEYL introduces gauge symmetry to describe electromagnetism and the theory of general relativity

1922 ● GEORGE POLYA rediscovers the 17 WALLPAPER GROUPS. The drawings in his paper describing these groups provide inspiration to the Dutch graphic artist M. C. ESCHER, who makes his first visit to the Alhambra this year, where he is inspired by the decorative tilings

1927 ● Henry Forder develops a new system of axioms for Euclidean geometry

1928 ● ABRAM BESICOVITCH proves that there is no smallest area swept out by a rod that is moved in a plane to have the net effect of a rotation of 180°. Zigzag motions of the rod can be designed to make the area arbitrarily small

● JAMES ALEXANDER defines the Alexander polynomial, the first knot invariant

1932 ● G. D. BIRKHOFF publishes a set of axioms for Euclidean geometry that incorporates the properties of the real numbers

● German mathematician Kurt Reidemeister describes the REIDEMEISTER MOVES and completes the classification of KNOTS with nine or fewer crossings

1935 ● Witold Hurewicz defines the higher-dimensional HOMOTOPY GROUPS.

1939 ● HEXAFLEXAGONS are invented by Arthur H. Stone, a graduate student at Princeton University

1942 ● Loran (long-range navigation), a system that uses radio signals and properties of HYPERBOLAS for determining the position of a ship or airplane, is invented at MIT

1949 ● BUCKMINSTER FULLER builds the first geodesic dome

1951 ● JEAN-PIERRE SERRE discovers the structure of the higher-dimensional HOMOTOPY GROUPS

1954 ● C. N. YANG and Robert L. Mills develop a non-Abelian GAUGE THEORY to describe isotopic-spin symmetry in physics

● RENÉ THOM develops COBORDISM theory

1956 ● JOHN MILNOR founds the study of DIFFERENTIAL TOPOLOGY with his discovery of the 28 EXOTIC SPHERES with different smooth structures

1957 ● The van Hiele model describing the levels of thinking through which students pass as they learn geometry is proposed by Pierre M. Van Hiele and Dina van Hiele-Geldorf. The levels in the model are visual, analysis, informal deduction, formal deduction, and rigor

1958 ● The School Mathematics Study Group (SMSG) develops a set of axioms for Euclidean geometry to be used in school mathematics

1961 ● EDWARD LORENZ discovers the phenomenon of CHAOS while using a computer to model a weather system. A very slight change in the initial conditions produces a great change in the output

1963 ● The Atiyah-Singer index theorem, which gives an important topological invariant of manifolds, is proved by MICHAEL ATIYAH and ISADORE SINGER. This result marks the beginning of the study of global analysis

1964 ● STEPHEN SMALE proves the POINCARÉ CONJECTURE for MANIFOLDS of dimension greater than four

1967 ● STEPHEN SMALE defines the HORSESHOE MAP, which demonstrates chaotic behavior

1968 ● JOHN CONWAY defines the Conway polynomial, a knot invariant

● French engineer PIERRE BEZIER creates BEZIER SPLINES as an aid in the design of automobiles

1970 ● JOHN CONWAY creates the game of Life, a computer model of population growth

1971 ● The Mariner 9 space mission uses Reed-Muller error-correcting codes based on AFFINE GEOMETRY to transmit photographs of the surface of Mars back to Earth

1972 ● P. M. de Wolff introduces SYMMETRY GROUPS of four-dimensional space to study the X-ray diffraction patterns of crystals

● RENÉ THOM begins the study of CATASTROPHE THEORY

1973 ● ROGER PENROSE discovers the PENROSE TILES, which can tile the plane only in a nonperiodic pattern

1974 ● The American lawyer Kenneth Perko discovers the equivalence of two knots in C. N. Little's table of knots that had been regarded as different for about 100 years. The two knots are known as the PERKO PAIR

1975 ● Steve Fisk proves the art gallery theorem, which says that the number of guards needed to survey a polygonal art gallery with n vertices is the largest integer that is less than or equal to $n/3$

1976 ● Using more than 1,000 hours of computer time, KENNETH APPEL and WOLFGANG HAKEN of the University of Illinois prove the four-color theorem, which says four colors are sufficient to color a map on a plane so that adjacent regions have different colors

1977 ● ROBERT CONNELLY constructs a closed, simply connected polyhedral surface that is not rigid, disproving a conjecture proposed by Euler in 1766 that all such surfaces must be rigid

● WILLIAM THURSTON conjectures that every three-dimensional manifold is composed of pieces, each having a geometric structure. This conjecture remains unproven in 2003

1978 ● At Brown University, THOMAS BANCHOFF and Charles Strauss use a computer to generate a film of slices and projections of a HYPERCUBE

● Michael Shamos writes a doctoral dissertation that is considered to be the foundation of COMPUTATIONAL GEOMETRY

1979 ● BENOIT MANDELBROT discovers the MANDELBROT SET

A quasicrystal
(P. C. Canfield and I. R.
Fisher/Ames Laboratory)

1982 ● Dan Shechtman produces the first QUASICRYSTAL, an alloy of aluminum and manganese that has X-ray diffraction patterns with fivefold ROTATIONAL SYMMETRY. Chemists and mathematicians still do not understand how atoms can be arranged to exhibit this kind of symmetry

● Michael H. Freedman proves the POINCARÉ CONJECTURE for four-dimensional MANIFOLDS

1983 ● JOHN CONWAY and Cameron Gordon prove that any planar representation of a complete GRAPH with more than seven vertices contains a subgraph that is the diagram of a nontrivial KNOT

● SIMON DONALDSON shows the existence of different smooth structures on four-dimensional Euclidean space

1985 ● Buckminsterfullerene, or the buckyball, a new form of highly symmetrical carbon, is discovered by Richard E. Smalley and Robert F. Curl Jr. of the United States and Sir Harold W. Kroto of Great Britain

● VAUGHN JONES discovers the Jones polynomial, a knot invariant related to statistical mechanics

● JEAN-MARIE LABORDE begins the development of Cabri Geometry, software for exploring properties of geometry objects and their relationships

1987 ● INGRID DAUBECHIES constructs the first family of WAVELETS defined on a finite interval

● The Geometer's Sketchpad dynamic geometry computer software is invented by Nick Jackiw

● James Gleick's book *Chaos: Making a New Science* is published and introduces the general public to CHAOS, FRACTALS, and DYNAMICAL SYSTEMS

1988 ● The nonexistence of a PROJECTIVE PLANE of ORDER 10 is proved by C. W. H. Lam using computer calculations

1989 ● Cameron Gordon and John Luecke prove that a KNOT is completely determined by the topological structure of its COMPLEMENT when embedded in a three-dimensional sphere

　　● RICHARD BORCHERDS proves the "monstrous moonshine theorem," which describes connections between the SYMMETRIES of a 196,833-dimensional shape, modular functions, and STRING THEORY

　　● Victor Vassiliev and Mikhail Goussarov independently discover a new class of knot invariants, called Vassiliev invariants

1991 ● Carolyn Gordon, David Webb, and Scott Wolpert prove that there are drums that are shaped differently but which nevertheless sound the same when struck. Their result says that "you can't hear the shape of a drum"

1993 ● JOHN CONWAY discovers a polyhedron that will fill space only in a nonperiodic fashion

David Webb and Carolyn Gordon with models of drums that sound alike
(Washington University in St. Louis)

　　● KRYSTYNA KUPERBERG constructs a smooth vector field on a three-dimensional manifold with no closed orbits, disproving a conjecture made by Herbert Seifert in 1950

　　● Peter Kronheimer and Tomasz Mrowka prove that the UNKNOTTING NUMBER of a (p, q)-TORUS KNOT is $(p – 1) (q – 1)/2$

1994 ● ALAIN CONNES introduces the study of NONCOMMUTATIVE GEOMETRY

　　● EDWARD WITTEN and Nathan Seiberg develop a pair of equations, known as the Seiberg-Witten equations, that can be used to compute the invariants of the four-dimensional MANIFOLDS studied by physicists

1995 ● Joel Haas and Roger Schlafly prove the double bubble theorem for bubbles of equal volume. The theorem says that the double bubble composed of pieces of three spheres meeting at an angle of 120° along a common circle has minimal surface area for the volume enclosed

1998 ● THOMAS HALES proves that, in the densest possible PACKING of identical spheres, the centers of the sphere form a FACE-CENTERED CUBIC LATTICE

1999 ● THOMAS HALES proves the honeycomb conjecture, which says that a hexagonal GRID is the best way to divide a plane into regions of equal area with the least total perimeter

● Yasumasa Kanada computes 206 billion decimal digits of pi

2000 ● Allan Wilks discovers an APOLLONIAN PACKING of a circle by circles with integer CURVATURES

● Jade Vinson discovers the first known HOLYHEDRON. It has 78,585,627 faces

● Michael Hutchings, FRANK MORGAN, Manuel Ritoré, and Antonio Ros prove the double bubble theorem for bubbles of unequal volume. The theorem says that the double bubble is composed of pieces of three spheres meeting at an angle of 120° along a common circle

● ROBERT CONNELLY, Eric Demaine, and Günther Rote prove that any polygonal framework, no matter how crinkled, can be unfolded in the plane into a CONVEX polygon

● Warwick Tucker of Cornell University shows that the differential equations given by EDWARD LORENZ in 1961 do, in fact, have a strange attractor

2001 ● Monica Hurdal begins using CONFORMAL mappings to give two-dimensional representations of the surface of the human brain

2002 ● MICHAEL FREEDMAN and Alexei Kitaev propose a topological quantum computer based on calculation of the JONES POLYNOMIAL

2003 ● Preliminary analysis from the NASA WMAP mission indicates that the geometry of the universe is either flat or slightly spherical

● Russian mathematician Dr. Grigori Perelman announces a proof of the Poincaré conjecture using methods from differential geometry

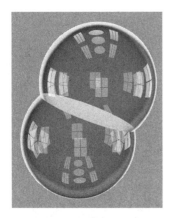

A double bubble (John M. Sullivan/University of Illinois)

SECTION FOUR
CHARTS & TABLES

Theorems

1 Two distinct points determine exactly one line and two distinct lines intersect in at most one point.
2 Vertical angles are congruent to each other.
3 Lines parallel to the same line are parallel to each other.
4 If two parallel lines are cut by a transversal, then each of the following is a pair of congruent angles: alternate interior angles, alternate exterior angles, and corresponding angles.
5 If two lines are cut by a transversal and alternate interior angles, alternate exterior angles, or corresponding angles are congruent to each other, then the two lines are parallel.
6 The base angles of an isosceles triangle are congruent.
7 If two angles of a triangle are congruent, the triangle is isosceles and the two congruent sides are opposite the congruent angles.
8 **Steiner-Lehmus Theorem** If two angle bisectors of a triangle are congruent, the triangle is isosceles.
9 In any triangle, the greater angle subtends the greater side and, conversely, the greater side subtends the greater angle.
10 **Triangle Inequality** The length of one side of a triangle is less than the sum of the lengths of the other two sides.
11 The sum of the measures of the interior angles of a triangle is 180°.
12 The measure of an exterior angle of a triangle is the sum of the measures of the two opposite interior angles.
13 **Pythagorean Theorem** In a right triangle, the square of the hypotenuse is equal to the sum of the squares of the other two sides.
14 **SAS** If two sides of one triangle are proportional to two sides of another triangle and the included angles are congruent, then the two triangles are similar.
15 **AA** If two angles of one triangle are congruent to two angles of another triangle, then the two triangles are similar.
16 **SSS** If corresponding sides of two triangles are proportional, then the two triangles are similar.
17 A line parallel to one side of a triangle cuts the other two sides proportionately and, conversely, if a line cuts two sides of a triangle proportionately, it is parallel to the third side.
18 The perpendicular bisectors of the three sides of a triangle intersect at a point, which is the circumcenter of the triangle.
19 The three angle bisectors of a triangle intersect at a point, which is the incenter of the triangle.

(continues)

Theorems *(continued)*

20 The three medians of a triangle intersect at a point, the centroid of the triangle.

21 The three altitudes of a triangle intersect at a point, called the orthocenter.

22 In an equilateral triangle, the circumcenter, the incenter, the centroid, and the orthocenter are the same.

23 The circumcenter, the centroid, and the orthocenter of a triangle are collinear and lie on the Euler line.

24 The centroid of a triangle trisects the segment joining the circumcenter and the orthocenter of the triangle.

25 The midpoints of the three sides of a triangle, the feet of the three altitudes of the triangle, and the midpoints of the segments joining the orthocenter of the triangle to the three vertices all lie on a circle, called the nine-point circle.

26 The center of the nine-point circle is the midpoint of the segment joining the orthocenter to the circumcenter. The radius of the nine-point circle is half the circumradius of the triangle.

27 **Feuerbach's Theorem** The nine-point circle of a triangle is tangent to the incircle of the triangle and to each of the three excircles of the triangle.

28 **Morley's Theorem** The three points of intersection of adjacent trisectors of the angles of a triangle form an equilateral triangle.

29 The diagonals of a parallelogram bisect each other.

30 Opposite angles of a parallelogram are congruent.

31 The diagonals of a rhombus are perpendicular.

32 **Varignon's Theorem** The four midpoints of the sides of a quadrilateral are the vertices of a parallelogram.

33 The measures of opposite angles of a convex cyclic quadrilateral sum to 180°.

34 **Ptolemy's Theorem** The product of the diagonals of a convex cyclic quadrilateral is equal to the sum of the products of opposite sides of the quadrilateral.

35 There is exactly one circle that passes through any three noncollinear points.

36 Congruent chords of a circle are equidistant from the center of the circle.

37 If a diameter of a circle bisects a chord that is not a diameter, it is perpendicular to the chord. Conversely, if a diameter of a circle is perpendicular to a chord, it bisects the chord.

38 A tangent to a circle is perpendicular to the radius that meets the tangent at the point of tangency.

39 The bisector of the angle between two tangents to a circle contains a diameter of the circle.

(continues)

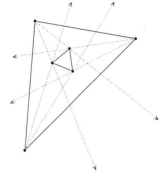

Morley's Theorem

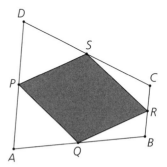

The Varignon parallelogram of quadrilateral *ABCD*

Theorems (*continued*)

40 If two parallel lines intersect a circle, the two arcs between the parallel lines are congruent.

41 An angle inscribed in a circle has half the measure of its intercepted arc.

42 The measure of the angle between two chords of a circle is one-half the sum of the intercepted arcs.

43 The angle between two secants to a circle that meet outside the circle is one-half the difference of the intercepted arcs.

44 The measure of an angle between a chord of a circle and a line tangent to the circle at an endpoint of the chord is one-half the measure of the intercepted arc.

45 If two chords of a circle intersect, the product of the lengths of the segments of one chord is equal to the product of the lengths of the segments of the other chord.

46 If a secant and a tangent to a circle meet at a point P, the secant intersects the circle at points S_1 and S_2, and the tangent meets the circle at a point T, then $PS_1 \cdot PS_2 = PT^2$.

47 A tangent to an ellipse forms congruent angles with the focal radii at the point of tangency.

48 The angle between a tangent to a parabola and the focal radius to the point of tangency is congruent to the angle between the tangent and the axis of the parabola.

49 The sum of the exterior angles of a polygon is $360°$. The sum of the interior angles of a polygon with n sides is $(n-2)180°$.

50 The ratio of the areas of two similar polygons is the square of the ratio of corresponding sides.

51 A regular polygon of n sides can be constructed with straightedge and compass if and only if the odd prime factors of n are distinct Fermat primes of the form $2^{2^k} + 1$ where k is an integer.

52 The diagonals of a regular pentagon divide each other in the golden ratio.

53 **Isoperimetric Theorem** Of all planar figures with a given perimeter, the circle has the greatest area. Of all figures with a given area, the circle has the smallest perimeter. Of all solids with a given surface area, the sphere has the greatest volume. Of all solids with a given volume, the sphere has the least surface area.

54 **Pick's Theorem** In a lattice whose fundamental parallelogram has unit area, the area of a simple polygon whose vertices are points of the lattice is $\frac{1}{2}p + q - 1$, where p is the number of lattice points on the boundary of the polygon and q is the number of lattice points inside the polygon.

55 **Euler's Theorem** Let A, B, C, and D be four collinear points. Then, using directed magnitudes, $AD \cdot BC + BD \cdot CA + CD \cdot AB = 0$.

(continues)

Theorems (continued)

56 **Theorem of Menelaus** Points L, M, and N on sides AB, BC, and CA or their extensions, respectively, of triangle ABC are collinear if and only if $\frac{AL}{LB}\frac{BM}{MC}\frac{CN}{NA} = -1$ using directed magnitudes.

57 **Ceva's Theorem** Let points L, M, and N be chosen on sides AB, BC, and CA, respectively, of triangle ABC. The lines AL, BM, and CN are concurrent if and only if $\frac{AL}{LB}\frac{BM}{MC}\frac{CN}{NA} = 1$ using directed magnitudes.

58 **The Butterfly Theorem** Let a complete quadrangle be inscribed in a circle. Each diagonal point of the quadrangle is the pole, under reciprocation with respect to the circle, of the line passing through the two other diagonal points.

59 The ratio of the volumes of two similar polyhedra is the cube of the ratio of corresponding sides.

60 A planar graph is a Schlegel diagram for a polyhedron if and only if the graph is connected and cannot be disconnected by the removal of any one or two vertices and their incident edges.

61 **Euler's Formula** For a simple closed polyhedron, the number of faces minus the number of edges plus the number of vertices is equal to 2.

62 In three-dimensional space, three distinct noncollinear points determine exactly one plane.

63 In three-dimensional space, two distinct planes are either parallel or intersect in a unique line.

64 **Arithmetic Mean/Geometric Mean Inequality** The geometric mean of two or more numbers is less than or equal to the arithmetic mean of those numbers.

65 **The Mohr-Mascheroni Construction Theorem** Any construction that can be made with compass and straightedge can be made by compass alone.

66 **The Poncelet-Steiner Construction Theorem** Any construction that can be made with compass and straightedge can be made with straightedge along with one given circle and its center.

67 Any construction that can be made with compass and straightedge can be made by folding paper.

68 The images of three collinear points under an isometry are collinear.

69 The images of parallel lines under an isometry are parallel.

70 An angle and its image under an isometry are congruent.

71 An isometry of the plane is uniquely determined from its action on a triangle.

72 An isometry of three-dimensional space is uniquely determined by its action on a tetrahedron.

73 If two triangles are congruent, there is an isometry that maps one to the other.

74 An isometry of the plane with exactly one fixed point is a rotation.

(continues)

Theorems (continued)

75 An isometry of the plane that has at least two points fixed is either the identity or a reflection across the line joining the fixed points.

76 A direct isometry of the plane is either a translation or a rotation. An opposite isometry of the plane is either a reflection or a glide reflection.

77 The composition of two translations is a translation.

78 The composition of two rotations is a rotation. The composition of two rotations about the same center is a rotation about that center.

79 The composition of two reflections across parallel lines is a translation. The composition of two reflections across intersecting lines is a rotation whose center is the point of intersection of the two lines.

80 Every isometry of the plane is either the identity or the composition of at most three reflections.

81 The collection of all isometries of a geometric space forms a group with composition as the binary operation.

82 A direct isometry of three-dimensional space is a translation, a rotation, or a twist. An opposite isometry of three-dimensional space is a reflection, a glide reflection, or a rotary reflection.

83 **Leonardo's Theorem** A finite group of isometries of the plane is either cyclic or dihedral.

84 A similarity of the plane is uniquely determined by its action on a triangle.

85 A triangle is similar to its image under a similarity.

86 A similarity without fixed points is an isometry.

87 A similarity that is not an isometry has exactly one fixed point.

88 A similarity of the plane with ratio of similitude different from one is the composition of a dilation with a rotation or a reflection.

89 If a similarity is an involution, it is a reflection, half-turn, or central inversion.

90 A direct similarity of the plane is either an isometry or a dilative rotation. An opposite similarity of the plane is either an isometry or a dilative reflection.

91 The image of a line under a dilation is a line and the image of a circle under a dilation is a circle.

92 An angle and its image under a dilation are congruent.

93 **Theorem of Pappus** If A, B, and C are three distinct points on a line and A', B', and C' are three distinct points on another line, then the three points of intersection of AB' with $A'B$, AC' with $A'C$, and BC' with $B'C$ are collinear.

94 **Theorem of Desargues** Two triangles that are perspective from a point are also perspective from a line.

95 Triangles that are perspective from a line are perspective from a point.

(continues)

Theorems *(continued)*

96 **Pascal's Theorem** The three points of intersection of pairs of opposite sides of a hexagon inscribed in a circle or other conic section are collinear.

97 **Brianchon's Theorem** The three diagonals connecting opposite vertices of a hexagon circumscribed about a circle or other conic section are concurrent.

98 A perspectivity between the points on two coplanar lines is uniquely determined by the images of two points.

99 **Fundamental Theorem of Projective Geometry** A projectivity between the points on two coplanar lines is uniquely determined by the images of three points.

100 The cross ratio of four collinear points is invariant under projection and, conversely, any transformation that preserves the cross ratio of every four collinear points is a projective transformation.

101 The image of a harmonic set of points under a projectivity is a harmonic set of points.

102 A projective plane of order q has $q^2 + q + 1$ points and $q^2 + q + 1$ lines.

103 An n-dimensional projective space of order q has $\dfrac{q^{n+1} - 1}{q - 1}$ points.

104 The intersection of two convex sets is convex.

105 **Hahn-Banach Separation Theorem** Every convex set in a Euclidean space is the intersection of half-spaces.

106 **Minkowski's Convex Body Theorem** An open, bounded, convex subset of the Cartesian plane that has point symmetry with respect to the origin and area greater than 4 must contain a point with integral coordinates that is different from the origin.

107 The boundary of a convex body in the plane is a simple closed curve.

108 **Barbier's Theorem** The perimeter of a plane convex body with constant width w is πw.

109 **Helly's Theorem** Let $N \geq n + 1$ and $n = 1, 2,$ or 3. If every $(n + 1)$ sets of a collection of N convex sets in n-dimensional space has nonempty intersection, then the intersection of all N sets is nonempty.

110 **Carathéodory's Theorem** Let S be a subset of n-dimensional Euclidean space and let P be in the convex hull of S. Then there exists a subset T of S such that T has at most $n + 1$ points and P is in the convex hull of T.

111 **Steinitz's Theorem** Let S be a subset of n-dimensional Euclidean space and let P be in the interior of the convex hull of S. Then there exists a subset T of S such that T has at most $2n$ points and P is in the interior of the convex hull of T.

112 In hyperbolic geometry, two triangles are congruent if and only if corresponding angles are congruent.

(continues)

Theorems *(continued)*

113 In hyperbolic geometry, the sum of the measures of the angles of a triangle is less than 180°.

114 In hyperbolic geometry, two triangles with the same angle sum have the same area.

115 In elliptic geometry, the sum of the measures of the angles of a triangle is greater than 180°.

116 In elliptic geometry, two triangles with the same angle sum have the same area.

117 **Four-Color Theorem** Using at most four colors, every map on a plane can be colored so that regions having a common boundary are colored differently.

118 **Jordan Curve Theorem** Every simple closed planar curve divides the plane into two parts.

119 **Brouwer Fixed Point Theorem** Every continuous map of a circular disk onto itself has at least one fixed point.

120 Any closed bounded surface is topologically equivalent to a sphere having zero or one crosscaps and any number, possibly zero, of handles.

121 **Gauss-Bonnet Theorem** The total curvature of a surface embedded in three-dimensional space is 2π times the Euler-Poincaré characteristic of the surface.

Euclid's Axioms

Postulates

Let the following be postulated:

1. To draw a straight line from any point to any point.
2. To produce a finite straight line continuously in a straight line.
3. To describe a circle with any center and distance.
4. That all right angles are equal to one another.
5. That, if a straight line falling on two straight lines makes the interior angles on the same side less than two right angles, the two straight lines, if produced indefinitely, meet on that side on which are the angles less than the two right angles.

Common Notions

1. Things that are equal to the same thing are also equal to one another.
2. If equals are added to equals, the wholes are equal.
3. If equals are subtracted from equals, the remainders are equal.
4. Things that coincide with one another are equal to one another.
5. The whole is greater than the part.

Straightedge and Compass Constructions

A straightedge or unmarked ruler is used to construct lines, rays, and segments. A compass is used to draw a circle with a given point as center and a given distance as radius.

Numbers in parentheses refer to other constructions in this list.

1　**Construct an equilateral triangle given one side.** At each endpoint of the side, draw a circle with radius equal to the side. A point of intersection of the two circles is the third vertex of the equilateral triangle; connect it to the two endpoints of the given side to obtain the required triangle.

2　**Construct the perpendicular bisector of a segment.** With the endpoints of the segment as centers, draw two congruent circles with radius greater than half the segment. Connect the two points of intersection of the circles; this line is the perpendicular bisector of the segment.

3　**Construct the midpoint of a segment.** Construct the perpendicular bisector (2) of the given segment, which will intersect the segment at its midpoint.

4　**Construct a line perpendicular to a given line at a given point on the line.** Draw a circle with center at the given point. Construct the perpendicular bisector (2) of the diameter connecting the points where this circle intersects the line; this is the required perpendicular.

5　**Drop a perpendicular from a given point to a given line.** Draw a circle with center at the given point and radius larger than the distance from the point to the line. Construct the perpendicular bisector (2) of the chord connecting the points where this circle intersects the line; this is the required perpendicular.

6　**Construct the reflection of a given point across a given mirror line.** Drop a perpendicular (5) from the point to the mirror line. Construct a circle that has center at the intersection of the perpendicular with the mirror line and passes through the original point. This circle intersects the perpendicular on the other side of the mirror line at the reflection of the given point.

7　**Construct a square given one side.** Construct a line perpendicular (4) to the given side at each endpoint. Draw two arcs on the same side of the given side, centered at the endpoints of the side and having radius equal to the side. The points of intersection of these arcs with the perpendiculars are the other two vertices of the required square.

8　**Make a copy of a segment on a given line with a given point as endpoint.** Draw a circle with center at the given point and radius congruent to the given segment. A radius of the circle lying on the given line is the required copy of the original segment.

(continues)

Straightedge and Compass Constructions *(continued)*

9 **Construct a segment whose length is the sum of the lengths of two given segments.** Let one of the two segments be *AB*. With endpoint *B* as center, draw a circle with radius congruent to the other segment. Point *A* and the point of intersection of this circle with the extension of *AB* are the two endpoints of the desired segment.

10 **Construct a segment whose length is the difference of the lengths of two segments.** Let the longer segment be *AB*. With endpoint *B* as center, draw a circle with radius congruent to the smaller segment. Point *A* and the point of intersection of this circle with segment *AB* are the two endpoints of the desired segment.

11 **Construct the bisector of a given angle.** With the vertex of the given angle as center, draw a circle and mark its points of intersection with the two sides of the angle. Draw two congruent circles with these two points as centers with radius large enough so that these two circles intersect. The bisector of the angle is the ray in the interior of the angle from the vertex of the angle to the intersection of these two circles.

12 **Make a copy of an angle.** Draw a circle at the vertex of the original angle and draw the segment whose endpoints are the intersection of this circle with the sides of the angle. Construct a second circle congruent to the first circle with center at the endpoint of a ray that will be one side of the copied angle. Construct a third circle with radius congruent to the segment and center at the point of intersection of the second circle with the ray. A ray from the new vertex to a point of intersection of the second and third circles will be the second side of the copied angle.

13 **Construct an angle whose measurement is the sum of two given angles.** Make a copy (12) of one of the angles. Make a copy (12) of the second angle on one of the sides of the first copy so that its second side is not in the interior of the first copy. The two rays not shared by the copied angles are the sides of the required angle.

14 **Construct an angle whose measurement is the difference of two given angles.** Make a copy (12) of the larger of the angles. Make a copy (12) of the smaller angle on one of the sides of the first copy so that its second side is in the interior of the first copy. The two rays not shared by the copied angles are the sides of the required angle.

15 **Construct a triangle, given one angle and its adjacent sides.** On one side of the given angle, make a copy (8) of one of the given sides with endpoint at the vertex; on the other side, make a copy (8) of the other given side with endpoint at the vertex. Connect the endpoints of these two segments to obtain the required triangle.

(continues)

Straightedge and Compass Constructions *(continued)*

16 **Construct a triangle, given two angles and the included side.** Make copies (12) of the two angles, one at each endpoint of the given side, so that the interiors of the two copies will intersect one another. The third vertex of the required triangle is the point of intersection of the two angle sides that do not lie on the given side of the triangle.

17 **Construct a triangle, given three sides.** Draw a circle centered at each endpoint of one of the sides, one with radius congruent to the second side and the other with radius congruent to the third side. One of the two points of intersection of these two circles will be the third vertex of the required triangle; connect it to the two endpoints of the first side to obtain the required triangle. This construction will work only if the sum of the lengths of every pair of sides is greater than the length of the other side.

18 **Construct a line parallel to a given line through a given point.** Drop a perpendicular (5) from the point to the line; construct a perpendicular (4) to this perpendicular at the given point. The second perpendicular will be parallel to the given line.

19 **Divide a given segment into any number n of congruent segments.** Let the given segment be AB. Construct a ray through point A that does not lie on segment AB. With a fixed radius, mark off n equally spaced points along this line, starting at the vertex A. Connect the last such point P to B. At each intermediate point, construct a parallel (18) to PB; these lines will divide AB into n congruent segments.

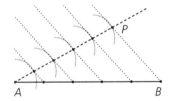

Division of the segment AB into five congruent segments

20 **For a triangle, construct the circumscribed circle.** Construct the perpendicular bisectors (2) of two sides of the triangle. Draw the circle that has center at the intersection of these two lines and that passes through any vertex of the triangle to obtain the circumscribed circle.

21 **For a triangle, construct the inscribed circle.** Construct the angle bisectors (11) of two angles of the triangle and drop a perpendicular (5) from their intersection to a side. Draw the circle that has center at the intersection of the two angle bisectors and passes through the foot of the perpendicular to obtain the inscribed circle.

22 **Construct a circle passing through three given points.** Construct the circumscribed circle of the triangle whose vertices are the three points (20).

23 **Construct a circle given an arc of the circle.** Mark three points on the arc, and then construct the circle through those three points (22).

24 **Locate the center of a given circle.** Construct the perpendicular bisectors (2) of any two chords of the circle; they intersect at the center of the circle.

25 **Construct an angle with measure 60°.** Draw a circle. Mark off an arc on the circle using its radius. Connect the center to each endpoint of the arc. The central angle thus constructed has measure 60°.

(continues)

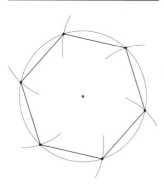

Construction of a regular hexagon

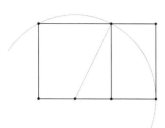

Construction of a golden rectangle

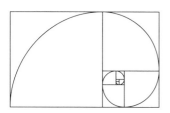

Construction of a golden spiral

Straightedge and Compass Constructions *(continued)*

26 **Construct an angle with measure 30°.** Construct an angle of measure 60° (25) and bisect it (11).

27 **Construct an angle with measure 15°.** Construct an angle of measure 30° (26) and bisect it (11).

28 **Construct a regular hexagon.** Draw a circle and choose a point on the circle. Using the radius of the circle, mark off six congruent arcs around the circumference. The chords subtending these arcs are the sides of a regular hexagon.

29 **Construct a golden rectangle.** Construct a square (7) and find the midpoint (3) of the base. Draw a circle with center at this midpoint that passes through a vertex on the opposite side of the square. Extend the base of the square until it intersects this circle; this extended base is the base of the golden rectangle. The altitude of the golden rectangle is a side of the original square.

30 **Construct a golden spiral.** Construct a square (7) on the altitude of a golden rectangle and draw a 90° arc in the square as shown. The remaining rectangle is also a golden rectangle; construct a square in this rectangle and draw a 90° arc as shown. Continue in this way; the spiral thus constructed will be a close approximation of a golden spiral.

31 **Construct the fourth proportional for three given segments.** Let three segments AB, CD, and EF be given. Extend segment AB and mark off segment BT congruent to CD so that B is between A and T. Make a copy (8) of segment EF with endpoints A and P that does not lie on the line AB. Connect B to P. Construct a parallel (18) to BP through T. Extend AP until it intersects the line through T at a point Q. Segment PQ is the fourth proportional since $\frac{AB}{CD} = \frac{EF}{PQ}$.

32 **Construct the arithmetic mean of two given segments.** Construct a segment (9) whose length is the sum of the lengths of the two segments. Find the midpoint (3) of this segment. The segment from one of the endpoints to this midpoint has length equal to the arithmetic mean of the lengths of the given segments.

33 **Construct the geometric mean of two given segments.** Construct a rectangle (45) whose width is congruent to one segment and whose altitude is congruent to the other. The side of a square (48) whose area is equal to this rectangle has length equal to the geometric mean of the lengths of the two segments.

34 **Construct a regular octagon.** Draw a circle. Draw a diameter and construct a diameter perpendicular (4) to it. Bisect two adjacent right angles (11) formed by the diameters and extend the angle bisectors so that the other two right angles are also bisected. Eight congruent arcs are formed in this way. The chords subtending these arcs are the sides of a regular octagon.

(continues)

Straightedge and Compass Constructions *(continued)*

35 **Construct a parallelogram given two sides and the included angle.** Make a copy (8) of the first of the two sides on a side of the given angle so that an endpoint coincides with its vertex and construct a parallel (18) to the other side of the angle through the other endpoint of this segment. Make a copy (8) of the second of the two sides on the other side of the given angle so that an endpoint coincides with its vertex and construct a parallel (18) to the first side of the angle through the other endpoint of this segment. The point of intersection of these two parallels is the fourth vertex of the parallelogram.

36 **Construct the tangent to a given circle at a given point on the circle.** Draw the radius to the given point and construct a perpendicular (4) to the radius at the given point. This line is the required tangent.

37 **Construct the tangent to a given circle from a given point in the exterior of the circle.** Construct the midpoint (3) of the segment whose endpoints are the given external point and the center of the given circle. Draw the circle centered at this midpoint with diameter equal to this segment. The line connecting the external point to either point of intersection of this circle with the given circle is the required tangent.

38 **Construct a common external tangent to two circles.** Construct the segment whose length is the difference (10) between the length of the larger radius and the length of the smaller radius. Draw a circle that is concentric with the larger circle and has radius congruent to this segment. Construct a tangent (37) to this new circle through the center of the smaller circle and draw the radius of the larger circle through the point of tangency. Construct the tangent (36) to the larger circle at the endpoint of this radius; this is the required tangent.

39 **Construct a common internal tangent to two circles.** Construct the segment whose length is the sum (9) of the length of the larger radius and the length of the smaller radius. Draw a circle that is concentric with the larger circle and has radius congruent to this segment. Construct a tangent (37) to this circle through the center of the smaller circle and draw the radius of the larger circle that passes through the point of tangency. The line tangent (36) to the larger circle at this point will also be tangent to the smaller circle.

40 **Construct an altitude of a polygon.** Drop a perpendicular (5) from a vertex to the base, extending the base if necessary. The segment connecting the vertex to the intersection of this perpendicular with the base is an altitude of the polygon.

(continues)

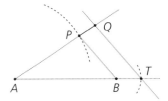

Construction of the fourth proportional of segments *AB*, *CD*, and *EF*

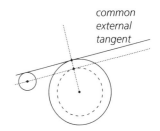

Construction of a common external tangent to two circles

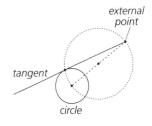

Construction of a tangent to a circle from an external point

Straightedge and Compass Constructions

CHARTS & TABLES

Straightedge and Compass Constructions *(continued)*

41 **Construct the orthocenter of a given triangle.** Drop a perpendicular (5) from each of two vertices to their opposite sides. The point of intersection of these two perpendiculars is the orthocenter of the given triangle. In some cases, the orthocenter may be outside the triangle.

42 **Construct a median of a triangle.** Join the midpoint (3) of a side to the opposite vertex.

43 **Construct the centroid of a given triangle.** The intersection of two medians (42) of the triangle is the centroid.

44 **Construct the nine-point circle belonging to a given triangle.** Construct the circle (22) passing through the midpoints (3) of the three sides of the triangle.

45 **Construct a rectangle given its base and altitude.** Construct perpendicular lines (4) at the two endpoints of the base of the rectangle. Draw two arcs on the same side of the base that have center at the endpoints of the base and have radius congruent to the altitude. The points of intersection of these arcs with the perpendiculars are the other two vertices of the rectangle.

46 **Construct a square whose area is the sum of the areas of two other squares.** Construct a point on the base *AB* of the larger square whose distance from vertex *A* is equal to the length of the side of the smaller square. Connect this point to the vertex adjacent to *A* and opposite the base. This segment is the side of the required square. Construct a square (7) on this segment.

47 **Construct a square whose area is the difference of the areas of two other squares.** Make a copy (8) of the side of the smaller square on the base *AB* of the larger square with one endpoint at vertex *A*. Construct a line perpendicular (4) to the base at the other endpoint of the copy. Draw a circle with center *A* and radius congruent to side *AB*. It will intersect the perpendicular at point *P*. The segment from point *P* to the base is the side of the required square. Construct a square (7) on this segment.

48 **Construct a square whose area is the same as the area of a given rectangle.** Construct a square (7) whose side has length equal to one-half (3) the difference (10) of the width and height of the rectangle and a square (7) whose side has length equal to one-half (3) the sum (9) of the width and height of the rectangle. The square (47) whose area is the difference of the areas of these two squares is the required square.

49 **Construct a vesica piscis.** Given a segment, draw two congruent circles with centers at the endpoints of the segment and radius equal to the segment. The vesica piscis is the region contained by both circles.

(continues)

Straightedge and Compass Constructions *(continued)*

50 **Construct a regular decagon.** Draw a radius of a circle with center O. Find the midpoint M (3) of the radius. Construct a tangent (36) to the circle at the endpoint of the radius. Draw a circle with center at the point of tangency passing through point M. Connect the point of intersection P of this circle with the tangent to the center O of the circle. Draw a circle with center P passing through the point of tangency; its intersection with segment OP is Q. Use a compass with radius set at OQ to mark off 10 congruent arcs along the circumference of the circle O. The chords subtending these arcs are the sides of a regular decagon.

51 **Construct a regular pentagon.** Connect alternate vertices of a regular decagon (50).

52 **Construct a regular pentagram.** Draw the diagonals of a regular pentagon (51).

53 **Construct a regular polygon with 15 sides.** Construct a regular decagon (50) in a circle. From one vertex, mark off an arc using the radius of the circle. The difference of the arc subtending the radius and the arc subtending the side of the decagon is an arc subtending the side of a regular polygon with 15 sides. Mark off 15 copies of this arc around the circle and draw in the chords subtending these arcs to obtain the required polygon.

54 **Construct a triangle with area equal to the area of a given quadrilateral.** Let the given quadrilateral be $ABCD$ with base AB. Connect vertex A to the opposite vertex C. Construct a parallel (18) to this segment through vertex D. Vertices B and C and the intersection E of this parallel with line AB are the vertices of the desired triangle.

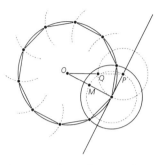

Construction of a regular decagon

Polygons

Triangles

For a triangle with base b, sides a and c, and height h:

perimeter $= a + b + c$

area $= \frac{1}{2}bh$

semiperimeter $= s = \frac{1}{2}(a + b + c)$

area $= \sqrt{s(s - a)(s - b)(s - c)}$

radius of incircle $= \dfrac{\sqrt{s(s - a)(s - b)(s - c)}}{s}$

radius of circumcircle $= \dfrac{abc}{4\sqrt{s(s - c)(s - b)(s - c)}}$

(continues)

Polygons *(continued)*

For an equilateral triangle with side a:

$$\text{area} = \frac{\sqrt{3}}{4}a^2$$

For a right triangle with altitude h on the hypotenuse dividing it into segments of length p and q:

$$h^2 = pq$$

Quadrilaterals

For a square of side a:

perimeter = $4a$

area = a^2

For a rectangle of sides a and b:

perimeter = $2a + 2b$

area = ab

For a parallelogram with bases b, sides a, and altitude h:

perimeter = $2a + 2b$

area = bh

For a trapezoid of height h with parallel bases b and B:

$$\text{area} = \frac{1}{2}(b + B)h$$

For a cyclic quadrilateral with sides a, b, c, and d:

$$\text{semiperimeter} = s = \frac{a + b + c + d}{2}$$

$$\text{area} = \sqrt{(s - a)(s - b)(s - c)(s - d)}$$

Regular Polygons

For a regular polygon with n sides of length a, apothem p, and radius r:

$$\text{interior angle} = \frac{180°(n - 2)}{n}$$

$$\text{exterior angle} = \frac{306°}{n}$$

perimeter = na

$$\text{area} = \frac{1}{2}npa = np^2 \tan\left(\frac{180}{n}\right) = \frac{1}{2}nr^2 \sin\left(\frac{360}{n}\right)$$

Polyhedra

For a rectangular parallelepiped with edges of lengths a, b, and c:

surface area $= 2(ab + ac + bc)$

volume $= abc$

For a prism of base area A and height h:

volume $= Ah$

For a pyramid with base area A and height h:

volume $= \frac{1}{3}Ah$

For the frustum of a pyramid with bases of areas b and B:

volume $= \frac{b + B + \sqrt{bB}}{3}h$

Regular Polyhedra (Part A)

	Number of Faces	Number of Edges	Number of Vertices	Surface Area	Volume
Tetrahedron	4	6	4	$\sqrt{3}\,e^2$	$\frac{\sqrt{2}}{12}e^2$
Cube	6	12	8	$6e^2$	e^3
Octahedron	8	12	6	$2\sqrt{3}\,e^2$	$\frac{\sqrt{2}}{3}e^3$
Dodecahedron	12	30	20	$3\sqrt{25 + 10\sqrt{5}}\,e^2$	$\frac{15 + 7\sqrt{5}}{4}e^3$
Icosahedron	20	30	12	$5\sqrt{3}\,e^2$	$\frac{5(3 + \sqrt{5})}{12}e^3$

Regular Polyhedra (Part B)

	Schläfli Symbol	Dihedral Angle Between Adjacent Faces	Number of Symmetries	Is the Polyhedron Rigid?
Tetrahedron	{3, 3}	70° 32′	24	yes
Cube	{4, 3}	90°	48	no
Octahedron	{3, 4}	109° 28′	48	yes
Dodecahedron	{5, 3}	116° 34′	120	no
Icosahedron	{3, 5}	138° 11′	120	yes

Curvilinear Figures and Solids

Circles

For a circle of radius r:

diameter $= 2r$

circumference $= 2\pi r$

area $= \pi r^2$

For an arc subtending an angle of measure θ:

arc length $= \dfrac{\theta}{180}\pi r$

For a sector cut off by angle of measure θ:

area $= \dfrac{\theta}{360}\pi r^2$

For a segment cut off by angle of measure θ:

area $= \dfrac{\theta}{360}\pi r^2 - \dfrac{1}{2}r^2\sin\theta$

Other Planar Curvilinear Figures

For an annulus with outer radius a and inner radius b:

area $= \pi(a^2 - b^2)$

For an ellipse with semi-axes a and b:

area $= \pi a b$

For a cycloid formed from a circle with radius r rolling along a line:

arc length of one arch $= 8r$

area under one arch $= 3\pi r^2$

Spheres

For a sphere of radius r:

surface area $= 4\pi r^2$

volume $= \dfrac{4}{3}\pi r^3$

For a spherical cap of height h on a sphere of radius r:

surface area $= 2\pi r h$

volume $= \dfrac{1}{3}\pi h^3(3r - h)$

Cylinders

For a right circular cylinder with radius r and height h:

lateral surface area $= 2\pi r h$

volume $= \pi r^2 h$

(continues)

Curvilinear Figures and Solids *(continued)*

Cones

For a right circular cone with radius r, slant height l, and height h:

lateral surface area $= \pi r \sqrt{r^2 + h^2} = \pi r l$

volume $= \frac{1}{3} \pi r^2 h$

For a frustum of a right circular cone with radii a and b, slant height l, and height h:

lateral surface are $= \pi(a + b)\sqrt{h^2 + (b - a)^2} = \pi(a + b)l$

volume $= \frac{1}{3} \pi h (a^2 + ab + b^2)$

Torus

For a torus with inner radius a and outer radius b:

surface area $= \pi^2 (b^2 - a^2)$

volume $= \frac{1}{3} \pi^2 (a + b)(b - a)^2$

Ellipsoid

For an ellipsoid with semi-axes a, b, and c:

volume $= \frac{4}{3} \pi abc$

Conic Sections

A conic section is the locus of a point P moving so that its distance from a fixed point, the focus, divided by its distance from a fixed line, the directrix, is a constant ε, the eccentricity.

	Circle	Ellipse	Parabola	Hyperbola
The intersection of a cone with . . .	A plane perpendicular to the axis of the cone	A plane that intersects the cone in a simple closed curve	A plane parallel to the generator of the cone	A plane that intersects both nappes of the cone
The locus of all points whose . . .	Distance from a fixed point, the center, is a constant	Distances from two fixed points, the foci, have a constant sum	Distance from a fixed point, the focus, is equal to its distance from a fixed point, the directrix	Distances from two fixed points, the foci, have a constant difference
Equation center $(0, 0)$	$x^2 + y^2 = r^2$	$\dfrac{x^2}{a^2} + \dfrac{y^2}{b^2} = 1$	$y = \dfrac{1}{4a}x^2$	$\dfrac{x^2}{a^2} - \dfrac{y^2}{b^2} = 1$
Equation center (x_0, y_0)	$(x - x_0)^2 + (y - y_0)^2 = r^2$	$\dfrac{(x - x_0)^2}{a^2} + \dfrac{(y - y_0)^2}{b^2} = 1$	$y - y_0 = \dfrac{(x - x_0)^2}{4a}$	$\dfrac{(x - x_0)^2}{a^2} - \dfrac{(y - y_0)^2}{b^2} = 1$
Comments	radius $= r$	If $a > b$, the major axis is horizontal. If $a < b$, the major axis is vertical. length of horizontal axis $= 2a$ length of vertical axis $= 2b$	If $a > 0$, the parabola opens upward. If $a < 0$, the parabola opens downward.	The hyperbola opens on the left and the right. length of horizontal axis $= 2a$ length of vertical axis $= 2b$
Focus	(x_0, y_0)	$(x_0 \pm \sqrt{a^2 + b^2}, y_0)$	$(x_0, y_0 + a)$	$(x_0 \pm \sqrt{a^2 + b^2}, y_0)$
Directrix		$x = x_0 \pm \dfrac{a^2}{\sqrt{a^2 + b^2}}$	$y = y_0 - a$	$x = x_0 \pm \dfrac{a^2}{\sqrt{a^2 + b^2}}$
Eccentricity	$\varepsilon = 0$	$\varepsilon < 1$ $\varepsilon = \dfrac{c}{a} = \dfrac{\sqrt{a^2 - b^2}}{a}$	$\varepsilon = 1$	$\varepsilon > 1$ $\varepsilon = \dfrac{c}{a} = \dfrac{\sqrt{a^2 + b^2}}{a}$
Asymptotes				$y - y_0 = \pm\dfrac{b}{a}(x - x_0)$
The second order equation $ax^2 + by^2 + cxy + dx + ey + f = 0$ gives this conic section if:	$a = b$	$\dfrac{a}{b} > 0$	$a = 0$ or $b = 0$	$\dfrac{a}{b} < 0$
Tangents and their properties	A tangent is perpendicular to a radius at the point of tangency.	A tangent makes equal angles with the rays from the point of tangency to the foci.	A tangent makes equal angles with the line from the focus to the point of tangency and a line perpendicular to the directrix through the point of tangency.	A tangent makes equal angles with the rays from the point of tangency to the foci.
Found in nature	Ripples in a pond	Planetary orbits	Trajectory of object thrown into the air	Trajectory of a comet that passes the sun only once
Applications	Wheels	Whispering gallery; medical treatment of kidney stones	Golden Gate Bridge; satellite dishes; solar oven	Loran; nuclear power cooling towers; hyperboloidal gears

Analytic Geometry

Plane Analytic Geometry

midpoint of the segment connecting points (x_1, y_1) and (x_2, y_2): $\left(\dfrac{x_1 + x_2}{2}, \dfrac{y_1 + y_2}{2}\right)$

distance between points (x_1, y_1) and (x_2, y_2): $\sqrt{(x_1 - x_2)^2 + (y_1 - y_2)^2}$

slope of the line passing through points (x_1, y_1) and (x_2, y_2): $\dfrac{y_2 - y_1}{x_2 - x_1}$

slope-intercept equation of the line with slope m and y-intercept $(0, b)$: $y = mx + b$

two-point equation of the line passing through points (x_1, y_1) and (x_2, y_2): $y = \dfrac{y_2 - y_1}{x_2 - x_1}(x - x_1) + y_1$

equation of the line passing through the point (x_1, y_1) with slope m: $y = mx + (y_1 - mx_1)$

The lines with slopes m_1 and m_2 are perpendicular if $m_1 m_2 = -1$.

distance from point (x_1, y_1) to line $Ax + By + C = 0$: $\left|\dfrac{Ax_1 + By_1 + C}{\sqrt{A^2 + B^2}}\right|$

tangent of the angle ψ between line $y = m_1x + b_1$ and $y = m_2x + b_2$: $\tan \psi = \dfrac{m_2 - m_1}{1 + m_1 m_2}$

centroid of the triangle with vertices (x_1, y_1), (x_2, y_2), and (x_3, y_3): $\left(\dfrac{x_1 + x_2 + x_3}{3}, \dfrac{y_1 + y_2 + y_3}{3}, \dfrac{z_1 + z_2 + z_3}{3}\right)$

area of the triangle with vertices (x_1, y_1), (x_2, y_2), and (x_3, y_3):

$\left|\dfrac{1}{2}\det\begin{bmatrix} x_1 & y_1 & 1 \\ x_2 & y_2 & 1 \\ x_3 & y_3 & 1 \end{bmatrix}\right| = \left|\dfrac{1}{2}(x_1y_2 + x_2y_3 + x_3y_1 - x_1y_3 - x_2y_1 - x_3y_2)\right|$

The point (x, y) in rectangular coordinates is the same as point (r, θ) in polar coordinates if :

$x = r \cos \theta$ and $y = r \sin \theta$ or
$r = \sqrt{x^2 + y^2}$ and $\theta = \tan^{-1}(y/x)$

Solid Analytic Geometry

distance between points (x_1, y_1, z_1) and (x_2, y_2, z_2): $\sqrt{(x_1 - x_2)^2 + (y_1 - y_2)^2 + (z_1 - z_2)^2}$

equation of the plane passing through points $(a, 0, 0)$, $(0, b, 0)$, and $(0, 0, c)$: $\dfrac{x}{a} + \dfrac{y}{b} + \dfrac{z}{c} = 1$

equation of the line passing through point (x_1, y_1, z_1) perpendicular to the plane

$Ax + By + Cz + D = 0$: $\dfrac{x - x_1}{A} = \dfrac{y - y_1}{B} = \dfrac{z - z_1}{C}$

distance from the point (x_1, y_1, z_1) to the plane $Ax + By + Cz + D = 0$: $\left|\dfrac{Ax_1 + By_1 + Cz_1 + D}{\sqrt{A^2 + B^2 + C^2}}\right|$

The point (x, y, z) in rectangular coordinates is the same as point (r, θ, z) in cylindrical coordinates if:

$x = r \cos \theta$, $y = r \sin \theta$, and $z = z$
$r = \sqrt{x^2 + y^2}$, $\theta = \tan^{-1}(y/x)$, and $z = z$

The point (x, y, z) in rectangular coordinates is the same as the point (r, θ, ϕ) in spherical coordinates if:

$x = r \sin \theta \cos \phi$, $y = r \sin \theta \cos \phi$, and $z = r \cos \theta$ or

$r = \sqrt{x^2 + y^2 + z^2}$, $\phi = \tan^{-1}(y/x)$, and $\phi = \cos^{-1}\left(\dfrac{z}{\sqrt{x^2 + y^2 + z^2}}\right)$

Trigonometry

All angles are given in degrees.

Trigonometric Functions

For a right triangle ABC with right angle C and sides a, b, and c opposite vertices A, B, and C

$$\text{sine of } \angle A = \sin A = \frac{\text{opposite}}{\text{hypotenuse}} = \frac{a}{c}$$

$$\text{cosine of } \angle A = \cos A = \frac{\text{adjacent}}{\text{hypotenuse}} = \frac{b}{c}$$

$$\text{tangent of } \angle A = \tan A = \frac{\text{opposite}}{\text{adjacent}} = \frac{a}{b}$$

$$\text{cotangent of } \angle A = \cot A = \frac{\text{adjacent}}{\text{opposite}} = \frac{b}{a}$$

$$\text{secant of } \angle A = \sec A = \frac{\text{hypotenuse}}{\text{adjacent}} = \frac{c}{b}$$

$$\text{cosecant of } \angle A = \csc A = \frac{\text{hypotenuse}}{\text{opposite}} = \frac{c}{a}$$

Pythagorean Identities

$\sin^2 A + \cos^2 A = 1$

$\tan^2 A + 1 = \sec^2 A$

$\sin^2 A + 1 = \csc^2 A$

Sum and Difference Formulas

$\sin(A + B) = \sin A \cos B + \cos A \sin B$ $\sin(A - B) = \sin A \cos B - \cos A \sin B$

$\cos(A + B) = \cos A \cos B - \sin A \sin B$ $\cos(A - B) = \cos A \cos B + \sin A \sin B$

$\tan(A + B) = \dfrac{\tan A + \tan B}{1 - \tan A \tan B}$ $\tan(A - B) = \dfrac{\tan A - \tan B}{1 + \tan A \tan B}$

$\cot(A + B) = \dfrac{\cot A \cot B - 1}{\cot A + \cot B}$ $\cot(A - B) = \dfrac{\cot A \cot B + 1}{\cot B - \cot A}$

Double Angle Formulas

$\sin 2A = 2 \sin A \cos A$

$\cos 2A = \cos^2 A - \sin^2 A = 1 - 2\sin^2 A = 2\cos^2 A - 1$

$\tan 2A = \dfrac{2 \tan A}{1 - \tan^2 A} = \dfrac{2}{\cot A - \tan A}$

Half Angle Formulas

$\sin \dfrac{A}{2} = \pm \sqrt{\dfrac{1 - \cos A}{2}}$

$\cos \dfrac{A}{2} = \pm \sqrt{\dfrac{1 + \cos A}{2}}$

(continues)

Trigonometry *(continued)*

$$\sin \frac{A}{2} = \pm \sqrt{\frac{1 - \cos A}{1 + \cos A}} = \frac{\sin A}{1 + \cos A} = \frac{1 - \cos A}{\sin A} = \csc A - \cot A$$

Signs depend on quadrant.

Power Formulas

$\sin^2 A = \frac{1}{2}(1 - \cos 2A)$

$\cos^2 A = \frac{1}{2}(1 + \cos 2A)$

Trigonometric Functions in the Cartesian Plane

For a directed angle θ with vertex at the origin (0,0) and initial side along the positive x-axis and terminal side passing through the point (x,y), the trigonometric functions are defined as follows, where $r = \sqrt{x^2 + y^2}$:

$\sin \theta = \frac{y}{r}$ $\csc \theta = \frac{r}{y}$

$\cos \theta = \frac{x}{r}$ $\sec \theta = \frac{r}{x}$

$\tan \theta = \frac{y}{x}$ $\cot \theta = \frac{x}{y}$

All Triangles

The following formulas are valid for any triangle ABC with sides a, b, and c opposite vertices A, B, and C:

Law of Sines

$$\frac{\sin A}{a} = \frac{\sin B}{b} = \frac{\sin C}{c}$$

Law of Cosines

$c^2 = a^2 + b^2 - 2ab \cos C$

Polar Form of Complex Numbers

For θ in radians, $r = \sqrt{x^2 + y^2}$, and $\theta = \tan^{-1}(y/x)$:

$x + iy = re^{i\theta} = r(\cos \theta + i\sin \theta)$

Spherical Trigonometry

In spherical trigonometry, an arc of a great circle is usually measured by the central angle that it subtends, rather than its length. The following formulas hold for a spherical triangle with vertices A, B, and C opposite sides a, b, and c, respectively, with each angle measure not more than 180°:

area of spherical triangle $ABC = \dfrac{(A + B + C - 180)\pi r^2}{180}$

Law of Sines

$$\frac{\sin A}{\sin a} = \frac{\sin B}{\sin b} = \frac{\sin C}{\sin c}$$

Cosine Law for Sides

$\cos a = \cos b \cos c + \sin b \sin c \cos A$

$\cos b = \cos c \cos a + \sin c \sin a \cos B$

$\cos c = \cos a \cos b + \sin a \sin b \cos C$

Cosine Law for Angles

$\cos A = -\cos B \cos C + \sin B \sin C \cos a$

$\cos B = -\cos C \cos A + \sin C \sin A \cos b$

$\cos C = -\cos A \cos B + \sin A \sin B \cos c$

Law of Tangents

$$\frac{\tan \frac{1}{2}(A + B)}{\tan \frac{1}{2}(A - B)} = \frac{\tan \frac{1}{2}(a + b)}{\tan \frac{1}{2}(a - b)}$$

Constants

$\sqrt{2}$	≈ 1.414213562
$\sqrt{3}$	≈ 1.732050808
π	≈ 3.141592654
e	≈ 2.718281828
1 radian	$\approx 57.29577951°$
$1°$	≈ 0.01745329252 radians
2π radians	$= 360°$

Greek Alphabet

A	α	alpha
B	β	beta
Γ	γ	gamma
Δ	δ	delta
E	$0, \varepsilon$	epsilon
Z	ζ	zeta
H	η	eta
Θ	θ, ϑ	theta
I	ι	iota
K	κ	kappa
Λ	λ	lambda
M	μ	mu
N	ν	nu
Ξ	ξ	xi
O	o	omicron
Π	π, ϖ	pi
P	ρ, Γ	rho
Σ	σ, ς	sigma
T	τ	tau
Y	υ	upsilon
Φ	ϕ, φ	phi
X	χ	chi
Ψ	ψ	psi
Ω	ω	omega

Abbreviations

A	area	**d**	diameter	**rect**	rectangle
AA	angle-angle	**ext**	exterior	**SA**	surface area
AAS	angle-angle-side	**int**	interior	**SAS**	side-angle-side
adj	adjacent	**LA**	lateral area	**sec**	secant
alt	altitude or alternate	**ln**	natural logarithm	**sin**	sine
ASA	angle-side-angle	**log**	logarithm for base 10	**st**	straight
B	base	**opp**	opposite	**supp**	supplementary
comp	complementary	**P**	perimeter	**TA**	total area
corr	corresponding	**pt**	point	**tan**	tangent
cos	cosine	**quad**	quadrilateral	**trap**	trapezoid
cot	cotangent	**r**	radius	**V**	volume
csc	cosecant	**rad**	radians	**vert**	vertical

Symbols

Geometry—Angles

$\angle$	Angle
$\angle A$	Angle with vertex A
$m\angle A$	The measure of angle with vertex A
$\perp$	Is perpendicular to
$\circ$	Degrees
$'$	Minutes
$''$	Seconds

Geometry—Circles

$\overset{\frown}{ACB}$	The arc that has endpoints A and B and contains point C
$\odot$	Circle
$\odot A$	Circle with center A
$\odot\overline{AB}$	Circle with center A and radius $\overline{AB}$
π	Pi, the ratio of the circumference of a circle to its diameter

Geometry—Lines

$\overleftrightarrow{AB}$	The line that passes through points A and B
$\overline{AB}$	The segment with endpoints A and B
$\overrightarrow{AB}$	The ray that has endpoint A and passes through B
$\parallel$ or $/\!/$	Is parallel to
$d(A, B)$	The distance from point A to point B

Geometry—Polygons and Polyhedra

$\triangle$	Triangle
$\square$	Parallelogram
$\square$	Rectangle
$\cong$	Is congruent to
$\not\cong$	Is not congruent to
$\approx$ or $\sim$	Is similar to
$\not\approx$	Is not similar to
$\{p\}$	A regular polygon with p sides
$\{p, q\}$	A regular polyhedron with faces that are regular polygons with p sides, q of which meet at each vertex

Vectors

$\mathbf{v}$ or $\vec{v}$	A vector
$\|\ \|$	Norm (of a vector)
$\cdot$	Dot product
$\times$	Cross product
$\otimes$	Tensor product

(continues)

Symbols *(continued)*

Other

dim(S)	The dimension of set S
$\overline{\overline{\wedge}}$	Is in perspective with respect to
H(AB, CD)	Point D is the harmonic conjugate of C with respect to A and B
H$_\infty$	The hyperplane at infinity

Relations

$=$	Equals
$\neq$	Is not equal to
$\approx$	Is approximately equal to
$::$	Equals (used for ratios)
$\propto$	Is directly proportional to
$\equiv$	Is congruent to
$\not\equiv$	Is not congruent to
$\approx$	Is isomorphic to
$\not\approx$	Is not isomorphic to

Numbers

e	The Euler number, the base of the natural logarithms
∞	Infinity

Logic

$\wedge$	And
$\vee$	Or
$\Rightarrow$ *or* $\rightarrow$	Implies
$\Leftrightarrow$ or $\leftrightarrow$	Is equivalent to; if and only if
$\neg$	Not
$/$	Not; used over any other symbol to negate its meaning
$\exists$	There exists
$\forall$	For all
$\therefore$	Therefore
$\because$	Since; because
$\ni$	Such that
$\blacksquare$ *or* $\square$	End of proof

Recommended Reading

Introductory Level

Abbott, Edwin Abbott. *Flatland.* Princeton, N.J.: Princeton University Press, 1991.

Banchoff, Thomas F. *Beyond the Third Dimension: Geometry, Computer Graphics, and Higher Dimensions.* New York: Scientific American Library, 1996.

Cederberg, Judith N. *A Course in Modern Geometries.* New York: Springer-Verlag, 2000.

Coxeter, H. S. M., and S. L. Greitzer. *Geometry Revisited.* New York: L. W. Singer, 1967.

Cromwell, Peter R. *Polyhedra.* Cambridge, U.K.: Cambridge University Press, 1999.

Crowe, Donald W., and Dorothy K. Washburn. *Symmetries of Culture: Theory and Practice of Plane Pattern Analysis.* Seattle: University of Washington Press, 1988.

Davis, Philip J. *Spirals: From Theodorus to Chaos.* Wellesley, MA: A K Peters, 1993.

Densmore, Dana, ed. Thomas L. Heath, trans. *Euclid's Elements.* Santa Fe, N.Mex.: Green Lion Press, 2002.

Devaney, Robert L. *Chaos, Fractals and Dynamics: Computer Experiments in Mathematics.* Reading, Mass.: Addison-Wesley, 1990.

Edmondson, Amy C. *A Fuller Explanation: The Synergetic Geometry of R. Buckminster Fuller.* Boston: Birkhäuser, 1987.

El-Said, Issam, and Ayse Parmen. *Geometric Concepts in Islamic Art.* Palo Alto, Calif.: Dale Seymour Publications, 1976.

Francis, Richard L. *The Mathematician's Coloring Book.* Lexington, Mass.: COMAP, 1989.

Friedrichs, K. O. *From Pythagoras to Einstein.* Washington, D.C.: MAA, 1965.

Gay, David. *Geometry by Discovery.* New York: John Wiley, 1998.

Gerdes, Paulus. *Geometry from Africa: Mathematical and Educational Explorations.* Washington, D.C.: MAA, 1999.

Hansen, Vagn Lundsgaard. *Geometry in Nature.* Wellesley, Mass.: A K Peters, 1994.

Hargittai, István, and Magdolna Hargittai. *Symmetry: A Unifying Concept.* Bolinas, Calif.: Shelter Publications, 1994.

Heath, Thomas L., ed. *The Thirteen Books of Euclid's Elements.* New York: Dover Publications, 1956.

Heilbron, J. L. *Geometry Civilized: History, Culture, and Technique.* Oxford, U.K.: Oxford University Press, 2000.

van Hiele, Pierre M. *Structure and Insight.* San Diego, Calif.: Academic Press, 1997.

Henderson, David W. *Experiencing Geometry in Euclidean, Spherical, and Hyperbolic Spaces.* Upper Saddle River, N.J.: Prentice Hall, 2001.

Honsberger, Ross. *Episodes in Nineteenth and Twentieth Century Euclidean Geometry.* Washington, D.C.: MAA, 1995.

Kappraff, Jay. *Connections: The Geometric Bridge between Art and Science.* New York: McGraw-Hill, 1991.

King, James. *Geometry through Circles with the Geometer's Sketchpad.* Emeryville, Calif.: Key Curriculum Press, 1996.

King, James, and Doris Schattschneider, eds. *Geometry Turned On: Dynamic Software in Learning, Teaching, and Research.* Washington, D.C.: MAA, 1997.

Krause, Eugene F. *Taxicab Geometry: An Adventure in Non-Euclidean Geometry.* New York: Dover Publications, 1986.

Lénárt, István. *Non-Euclidean Adventures on the Lénárt Sphere.* Emeryville, Calif.: Key Curriculum Press, 1996.

Martin, George E. *Transformation Geometry.* New York: Springer-Verlag, 1982.

McLeay, Heather. *The Knots Puzzle Book.* Emeryville, Calif.: Key Curriculum Press, 2000.

Meyer, Walter. *Geometry and Its Applications.* San Diego, Calif.: Academic Press, 1999.

Muller, Jim. *The Great Logo Adventure.* Austin, Tex.: Doone Publications, 1997.

Olds, C. D., Anneli Lax, and Giuliana Davidoff. *The Geometry of Numbers.* Washington, D.C.: MAA, 2000.

Pedoe, Dan. *Circles: A Mathematical View.* Washington, D.C.: MAA, 1995.

———. *Geometry and the Visual Arts.* New York: Dover Publications, 1983.

Posamentier, Alfred. *Advanced Euclidean Geometry: Excursions for Students and Teachers.* Emeryville, Calif.: Key Curriculum Press, 2002.

Posamentier, Alfred, and William Wernick. *Advanced Geometric Constructions.* Palo Alto, Calif.: Dale Seymour Publications, 1988.

Ranucci, E. R., and J. E. Teeters. *Creating Escher-Type Drawings.* Palo Alto, Calif.: Creative Publications, 1977.

Schattschneider, Doris. *Visions of Symmetry: Notebooks, Periodic Drawings, and Related Works of M. C. Escher.* New York: W. H. Freeman, 1990.

Senechal, Marjorie, and George Fleck. *Shaping Space.* Boston: Birkhäuser, 1988.

Serra, Michael. *Discovering Geometry: An Inductive Approach.* Emeryville, Calif.: Key Curriculum Press, 1997.

Smart, James R. *Modern Geometries.* Pacific Grove, Calif.: Brooks/Cole, 1998.

Thompson, D'Arcy W. *On Growth and Form.* New York: Dover Publications, 1992.

Trudeau, Richard J. *The Non-Euclidean Revolution.* Boston: Birkhäuser, 1987.

de Villiers, Michael. *Rethinking Proof with The Geometer's Sketchpad.*
    Emeryville, Calif.: Key Curriculum Press, 1999.
Weeks, Jeffrey. *Exploring the Shape of Space.* Emeryville, Calif.: Key
    Curriculum Press, 2001.

Videos, Software, and Websites

Amenta, Nina. *Kali.* Software for Macintosh and Windows. Available online.
    URL:http://humber.northnet.org/weeks/.
Clarke, Arthur C. *Fractals: The Colors of Infinity.* Princeton, N.J.: Films for the
    Humanities and Sciences, 1997. VHS. 30 minutes.
Devaney, Robert L. *Professor Devaney Explains the Fractal Geometry of the
    Mandelbrot Set.* Emeryville, Calif.: Key College Publishing. VHS. 70 minutes.
———. "The Geometry Junkyard." Available online. URL:
    http://www1.ics.uci.edu/~eppstein/junkyard/.
The Geometry Center. *Not Knot.* Wellesley, Mass.: A K Peters, 1991. VHS. 16
    minutes.
———. *Outside In.* Wellesley, Mass.: A K Peters, 1994. VHS. 22 minutes.
———. *KaleidoTile.* Minneapolis: The Geometry Center. Software for
    Macintosh. Available online. URL: http://humber.northnet.org/weeks/.
Jackiw, Nick. *The Geometer's Sketchpad Dynamic Geometry Software.*
    Emeryville, Calif.: Key Curriculum Press, 2001. Software for Macintosh and
    Windows.
Laborde, Jean-Marie, and F. Bellemain. *Cabri Geometry II.* Dallas, Tex.: Texas
    Instruments, 1994. Software for Macintosh and Windows.
Lee, Kevin. *KaleidoMania!* Emeryville, Calif.: Key Curriculum Press, 2000.
    Software for Macintosh and Windows.
Weeks, Jeffrey. *The Shape of Space.* Emeryville, Calif.: Key College
    Publishing, 2000. VHS. 20 minutes.
Weisstein, Eric. *World of Mathematics.* Available online. URL:
    http://mathworld.wolfram.com/.

Classroom Resources

Frame, Michael, and Benoit B. Mandelbrot. *Fractals, Graphics, and
    Mathematics Education.* Washington, D.C.: MAA, 2002.
Lindquist, Mary Montgomery, and Albert P. Shulte, eds. *Learning and Teaching
    Geometry, K-12.* Reston, Va.: NCTM, 1987.
Mammana, Carnelo, and Vinicio Villani, eds. *Perspectives on the Teaching of
    Geometry for the 21st Century: ICMI Study.* Dordrecht, Netherlands:
    Kluwer, 1998.

National Council of Teachers of Mathematics. *Curriculum and Evaluation Standards for School Mathematics.* Reston, Va.: NCTM, 2000.

———. *Navigating through Geometry in Grades 3–5.* Reston, Va.: NCTM, 2001.

———. *Navigating through Geometry in Grades 6–8.* Reston, Va.: NCTM, 2001.

———. *Navigating through Geometry in Grades 9–12.* Reston, Va.: NCTM, 2001.

———. *Navigating through Geometry in Prekindergarten–Grade 2.* Reston, Va.: NCTM, 2001.

Advanced Level

Casey, James. *Exploring Curvature.* Braunschweig/Wiesbaden, Germany: Vieweg, 1996.

Coxeter, H. S. M. *Introduction to Geometry.* New York: John Wiley, 1989.

———. *Non-Euclidean Geometry.* Washington, D.C.: MAA, 1998.

———. *Projective Geometry.* New York: Springer-Verlag, 1994.

———. *Regular Polytopes.* New York: Dover Publications, 1973.

Gorini, Catherine A., ed. *Geometry at Work: Papers in Applied Geometry.* Washington, D.C.: MAA, 2000.

Graver, Jack. *Counting on Frameworks.* Washington, D.C.: MAA, 2001.

Grünbaum, Branko, and G. C. Shephard. *Tilings and Patterns.* New York: W. H. Freeman, 1989.

Henderson, David W. *Differential Geometry: A Geometric Introduction.* Upper Saddle River, N.J.: Prentice Hall, 1998.

Hilbert, David, and S. Cohn-Vossen. *Geometry and the Imagination.* New York: Chelsea, 1999.

Livingston, Charles. *Knot Theory.* Washington, D.C.: MAA, 1993.

Mandelbrot, Benoit. *The Fractal Geometry of Nature.* New York: W. H. Freeman, 1983.

Okabe, Atsuyuki. *Spatial Tessellations: Concepts and Applications of Voronoi Diagrams.* New York: John Wiley, 2000.

O'Rourke, Joseph. *Computational Geometry in C.* Cambridge, U.K.: Cambridge University Press, 1996.

Radin, Charles. *Miles of Tiles.* Providence, R.I.: AMS, 1999.

Senechal, Marjorie. *Quasicrystals and Geometry.* Cambridge, U.K.: Cambridge University Press, 1995.

Stahl, Saul. *The Poincaré Half-Plane.* Boston: Jones and Bartlett, 1993.